DENYING SCIENCE

REFLECTIONS ON THOSE WHO REFUSE TO ACCEPT THE RESULTS OF SCIENTIFIC STUDIES

PASCAL de CAPRARIIS

authorHOUSE®

AuthorHouse™
1663 Liberty Drive
Bloomington, IN 47403
www.authorhouse.com
Phone: 1 (800) 839-8640

Published by AuthorHouse   06/30/2017

ISBN: 978-1-5246-9807-2 (sc)
ISBN: 978-1-5246-9806-5 (e)

For Nancy

"Age cannot wither her, nor custom stale her infinite variety."
Shakespeare

CONTENTS

PREFACE

Ever since the Reagan administration, the space separating scientists and non-scientists has widened. At one time people accepted the technologies that developed from scientific studies, but expressed no opinions about the basic science because, they often would say, they were not "science people." But a not too subtle change has occurred since the 1980s, in that people still are willing to admit that they know nothing about science, but many of them do not hesitate to criticize the results of scientific studies. I am thinking particularly of environmental regulations intended to protect the public. Someone with no knowledge of the details of a topic such as air pollution feels comfortable complaining about federal regulations restricting the levels of emissions from industrial sources. Others go to school board meetings and ask (or demand) to have materials based on their religious beliefs replace the material on Evolution in the Biology curriculum. And others routinely deny the relevance of data showing that carbon emissions are trapping heat in the atmosphere, causing global warming. The most vocal of the "climate-change" deniers are often politicians and media pundits, none of whom have any background in science.

The denial of scientific knowledge has become respectable, to the point that it threatens more than the education of children; it threatens the health of the public and the stability of our society as climatic changes due to global warming occur. So, the growth of "science denial," goes much further than just a lack of interest in learning about the world about us. Today, some people go out of their way to try to convince legislators and the general public that scientists are not doing objective, reliable work, but instead, are "crying wolf" about imagined problems in order to get grants to subsidize further studies. The efforts to discredit scientific studies are funded by the fossil fuel industry, and are facilitated by the existence of social media, which enable anyone to say nearly anything without being asked for evidence or for their qualifications.

The topics covered in this book highlight some of the differences between reputable studies and the public relations efforts of groups that reject the studies in order to advance their own agendas.

INTRODUCTION

A simple statement from a distinguished scientist (albeit one whom few people have heard of) provides insight into the discipline we call science.

"Science is not a body of facts. Science is a method for deciding whether what we choose to believe has a basis in the laws of nature or not."

That statement was made by Marcia McNutt, a geophysicist who once was the director of the U.S. Geological Survey, and is now the President of the National Academy of Sciences.

Think about what she said. Science provides a way to decide whether we should believe something. The criterion is simple. If the "thing" does not agree with basic laws of nature, there is no point in considering it any further, so it should be rejected. On the other hand, just because it may be consistent with the laws of nature does not mean it is correct: just that we cannot automatically reject it. At least not yet. That is the basis for the statement scientists often make, that scientific knowledge is provisional. "Provisional" does *not* mean that the knowledge is uncertain or questionable or debatable. It means that it may be shown to be wrong someday, but until then it is sensible to accept it.

Of course, scientists are human, and sometimes they find it hard to be consistent. Like everyone else, they are subject to confirmation bias, so they may subconsciously select only evidence that fits their idea of how something works. And they may be reluctant to reject something even if it is refuted by data. John Steinbeck wrote about this tendency once in his book *The Log From The Sea Of Cortez* (an underground classic which is well worth reading). He said:

"There is one great difficulty with a good hypothesis. When it is completed and rounded, the corners smooth and the content cohesive and coherent, it is likely to become a thing in itself, a work of art. It is then like a finished sonnet or a painting completed. One hates to disturb it. Even if subsequent information should shoot a hole in it, one hates to tear it down because once it was beautiful and whole."

But there is not much danger in this kind of wishful thinking, because when an idea is published, others who think that the results are not correct will immediately try to reproduce them. And if they cannot come up with the same results, they will be sure to announce the fact. This kind of competition ensures that provisional knowledge is either validated or refuted before too long. Remember Cold Fusion? It took very little time before it was discredited.

Of course, some people scoff at the words "provisional knowledge," but that is because they do not understand how science progresses, or because they do not want to accept a recommendation from a scientific study. Unfortunately, both attitudes are common.

Another way to think about scientific progress was provided by Jerome Groopman, a distinguished Oncologist: "Science is the accretion of provisional certainties."

We test hypotheses and retain the ones that pass the tests, so knowledge of a subject grows by accretion.

Conflicting Belief Systems

There are two important belief systems that often conflict with science. They are important because they represent the foundations of thinking processes that are common in society. They are religion and capitalism. They can conflict with science because people will sometimes reject scientific findings that either appear to contradict some religious belief or, because taking the recommendations based on science seriously might require major adjustments in the nation's economy. In both cases, when people reject the results of a scientific study, they are rejecting the credibility of the "output" of the scientific method.

Recall Dr. McNutt's definition of science. We try to determine if an observation or some hypothesis about how a process works conforms to the basic laws of nature. If the conclusion is positive, we assert that "this is true to the extent that it is consistent with what we know about how natural systems work."

To use one example, someone who rejects the conclusion that the Earth is a few billion years old because the stories in the Bible imply that it cannot be older than several thousand years, is rejecting more than laboratory tests: that person is rejecting a major branch of physics. You are perfectly free to reject modern science in favor of Bronze Age mythology, but you should at least be consistent. If the principles of nuclear physics that underlie radiometric dating are wrong, then many aspects of our technological society could not function. Nuclear power plants come to mind, as do radiation treatments for cancer. If you reject one application of nuclear fission, to be logical, you should reject many more. You cannot cherry-pick the applications to suit your predilections and expect to be taken seriously.

To use another example, when politicians and corporate executives balk at the thought of curtailing the use of fossil fuels to maintain the planet's average temperature at levels that will not trigger major climatic disasters, they are basing their objections on economic and political considerations, not on an understanding of how fluids such as the atmosphere and the oceans react to radiative heating. They prefer not to be inconvenienced by an economic downturn caused by a shift away from the use of fossil fuels. They would rather condemn future generations to extremely harsh climatic conditions, ones that will completely disrupt the economies of the entire planet, and force mass migrations of billions of people and result in the extinction of numerous groups of animals and plants.

Unfortunately for those who reject science, the laws of nature trump stories based on Bronze Age mythology and those based on economic theories, that in addition to having little to no supporting evidence, are completely independent of anything related to natural processes.

So let's get on with it. We will start by discussing how science and religion interact in our society today.

SCIENCE AND RELIGION

There should be no controversies between science and religion because in principle, they represent completely different ways to think about the world. Briefly, science is a way to ask why inanimate things behave the way they do (Newton's laws and Maxwell's equations come to mind as reasons), and religion is a way to remind animate things (people) about how they should interact with each other (The Sermon on the Mount comes to mind as guidance).

The subtexts of the two disciplines also differ. Scientists are aware that what seems to be a good explanation of a phenomenon may turn out to be oversimplified, so they try to find ways to improve it, or to replace it with something better. That is how science works and that is why scientific knowledge has progressed so much in the last few hundred years. On the other hand, religion has not progressed very much since the Bronze Age, because religion claims to have all the answers – therefore nothing needs to change. So, one discipline is trying to learn more and the other is saying that what it is now, and has always been, is sufficient. In one case, inquiry is the means to an end, which is knowledge. In the other case, ministry is the means to an end, which is salvation.

The means and goals of these two "enterprises" are completely different, yet controversies often occur because of encroachments. That is, some people will not stay on their side of the line. The most common encroachment occurs when someone claims that the results of a scientific study must be wrong because they contradict something in the Bible. The Theory of Evolution is the one I see mentioned in the media fairly often. It is taught in public school Biology classes because the Biology profession regards it as the foundation on which all biological knowledge rests.

That is a pretty strong endorsement, but it is not enough for some people, even though very few of those who object to teaching Evolution in public school Biology classes are scientists, and even fewer of them are Biologists or Geologists, so they have no credentials that might suggest they know anything about the subject. But they *know* Evolution is wrong because the Bible says that all living organisms were created during the six days of creation described in the Book of Genesis.

I said earlier that religion claims to have all the answers. There is no doubt. There is no uncertainty. What has always been assumed to be true, will always be assumed to be true. No scientist has that kind of confidence. But of course, no scientist depends on divine revelation for knowledge about how the world works. Scientists must work to learn things. Nothing is handed to them.

So we have divine revelation versus focused inquiry. They are two completely incompatible sources of knowledge. How can what one learns from these two sources be reconciled? I cannot answer that question, but I majored in science at a Jesuit university, at which many of the science faculty were priests who had no difficulty answering it for themselves. On the other hand, most of the most ardent opponents of teaching Evolution in the public schools are not scientists, so even if they are ordained ministers, they are arguing against a position they know nothing about. For that reason, it is not surprising that their objections are of no concern to so many scientific members of religious orders.

Instead of listening to those who criticize the science taught in the public schools, a more fruitful way to spend one's time is to ask questions about the beliefs of those who object to the standard curriculum.

Societal Implications of Unquestioned Religious Beliefs

One of the things Susan Jacoby discusses in her book, *The Age of American Unreason*, are demands by the religious community that go beyond the guarantee of religious freedom. Freedom of religion used to mean the right to worship as one pleases. But many people now believe that it also means that their beliefs and practices should be immune from the scrutiny applied to other social institutions. That is, they feel that their beliefs deserve respect to the point that they should not be subject to criticism; that they should be allowed to say or do whatever they wish, as long as what they wish to say or do is based on "deeply held religious beliefs."

Jacoby wrote that this kind of thinking leads to mindless tolerance, because "it puts observable scientific facts, which are subject to proof, on the same level as what amounts to unprovable supernatural fantasy." That equivalence allows Creationists to claim that the Earth cannot be 4.5 billion years old, because adding up the ages of men in the Old Testament would have it be about 6,000 years old. And that age allows them to reject the evolutionary development of organisms because 6,000 years would not provide enough time for evolutionary changes to occur. In addition, the Old Testament clearly states that God created every "kind" of thing just once, in the Garden of Eden, which contradicts the evolution of species by natural selection. So they would have us replace modern scientific findings with Bronze Age mythology.

Biblical Literalists, usually called Creationists, are asking us to accept their belief that the stories in the Old Testament are literally true, and to reject any ideas that are not compatible with them. That is what Jacoby means by "mindless tolerance." They want us to replace modern science with what they call Creation Science, but for which "ersatz science" is more appropriate.

But there is another problem with this way of thinking: it leads to Relativism. If ideas related to the beliefs of one religion warrant respect, can we refuse to show the same respect to the beliefs of other religions? If all are deserving of the same degree of tolerance (as they must be if we are to avoid the "Establishment" problem in the Constitution), we quickly start sliding down the slippery slope which leads us to the conclusion that all religions are equivalent, a position called Syncretism, that few ministers or priests would condone. After all, if your religion's beliefs must be put on the same pedestal as scientific findings, the beliefs of other religions should also be treated in the same way. For example, what about Animists, who believe that there are spirits associated with the trees in the forest? Should we agree that claims about tree spirits are as valid as the ones made about the Rapture? Or should we say that the claims about the Rapture are as valid as those made about tree spirits? Where does it stop?

We don't have that problem with science. When two different explanations for a phenomenon exist, they are tested to see which one provides a better, more accurate explanation for whatever is observed. And if neither seems to be adequate, some different explanation will be suggested, until everyone is satisfied that a good one is at hand. We don't have to deal with the ambiguity that religions are subject to.

The criteria for settling scientific disputes do not have counterparts for religious disputes, because no tests are available to let us decide unequivocally, between different claims. That complicates matters when different sects expect their views to be accepted. So, a society needs a way to prevent disputes from causing civil unrest. Without one, the following scenario will be common.

If your God is the omniscient, omnipotent, jealous being you believe Him to be, and if He is the only true God, then the ones worshipped by other people must clearly be figments of their imaginations. In that case, it makes sense for you to tell other people about Him, and argue that they need to worship only Him.

But what if the guy standing next to you believes *his* God is the only omniscient, omnipotent, jealous being, and therefore you should be worshipping *that* God instead of the obviously fake one you grew up worshipping? And if a third person in the crowd says that *his* God...etc., etc. Pandemonium could develop very quickly.

The problem of course is that they cannot *all* be right about their gods, so how is anyone else in the crowd going to be able to decide which god is the "true" God?

Unfortunately, "true believers" rarely are rational enough to admit that their god is no more credible than anyone else's, and that is the reason for the "Establishment" clause in the First Amendment to our Constitution. It enables you to worship as you please, which means that it protects you from being forced to worship someone else's god. After all, history suggests that without the protection provided by that clause in the amendment, our society would have a god imposed on it by the government.

Those who contend that religion is needed because science does not provide enough do not understand that scientists share the belief that they do not have all the answers, which is why they are always looking for ways to expand their understanding, *but only by means of verifiable statements.*

Random Comments on These Subjects

The next dozen or two pages contain items selected from a blog I created to allow me to express my opinions on a variety of topics. I selected some of the ones dealing with science and religion that are general enough as not to be tied to the newspaper headlines at the time they were written. Each item is separated from its neighbors by several lines to indicate when one ends and another begins. You will notice some overlap, because these items were written several months to a year or two apart, and I did not expect readers of the blog to remember every detail of every post. I have not "corrected" this problem here, because I feel that a little repetition helps convert short-term memory into long-term memory, and remembering what you read is the whole point of reading, isn't it? So here we go.

In one of his essays, Umberto Eco said that scientific studies provide reliable sources of information that are correct enough to base decisions on. "Correct enough" are important words, because as he put it, although we can all agree that dolphins are mammals, and that the water molecule has two hydrogen molecules and one oxygen molecule, the scientific explanations of how some phenomena work are revised from time to time.

Scientific explanations are subject to revision, as more information is obtained, and when such topics are revised, most people are likely to accept the revisions. No one starts a war over the details of how fluids behave as they pass over a solid surface or how atoms behave at extremely low temperatures. And because such explanations are subject to revision, we should not be surprised if some of them are dropped from the "known" category with time, because they are no longer considered good explanations. New information can "falsify" the old.

Eco's point (following Karl Popper) was that it is important to distinguish between things that can be "falsified" as new information becomes available, and those which cannot be falsified. In this context, "falsified" means that they are recognized to be incorrect because better data or a better explanation has become available. If an explanation cannot be improved, we are not doing science. Although things that can be falsified cannot be accepted as "true" in the sense that religious adherents demand of their beliefs, they allow us to function as a society, and quite often make our lives easier, as technologies are developed and perfected. So, although falsifiable information must be considered to be provisional, because it is subject to revision, that is no reason to reject this kind of information, because as revisions are made, it is improving.

A good example is provided by airplanes. The details of how air flows around an airplane's wings are subject to revision, because turbulent flow is very complicated. Over the years, with the development of larger and faster computers, the principles governing the flow have emerged, albeit slowly, but even today, scientists do not claim that they have a complete understanding of the phenomenon (far from it). But few people use that as a reason not to fly. "Provisional" can be good enough.

On the other hand, if what is known in some discipline cannot be made more precise (which is equivalent to saying it is completely understood, down to the last detail), it cannot be falsified. But how can we be sure that we know everything about the discipline? What evidence exists that anything we know is perfect?

I have gone into all this, of course, to make a point about Creationists. They call what they do "Creation Science," but their goal is not a better understanding of how some physical system works. Their goal is to find a way to relate some physical phenomenon to one or more passages in the Bible. But that is not what scientists do. Scientists want to know more, whereas Creationists want to shackle knowledge to the Bible.

Creationists do not really care about learning something new: they are only interested in finding ways to verify statements found in the Bible. The reason they call what they do science is because they want their ideas taught in science classes in the public school systems, something that would violate the "establishment" clause in the First Amendment, because it involves religion, not science.

They are welcome to believe that a book written down in the seventh century BCE, based on Bronze Age tales going back to the third millennium contains valid information about the origin of the universe. But no one who does not share their religious beliefs agrees with them, and if religion is the only support for their beliefs, it is not appropriate to have those beliefs taught as science with the imprimatur of the government.

A Biologist at the Univ. of Washington once published an Op-Ed article in *The New York Times* in which he said that he starts his course with "The Talk." He said that he tells his class that the Theory of Evolution is the foundation of modern Biology, and that without Evolution, nothing in Biology makes sense.

This approach differs a bit from that taken by many Biologists and Geologists, who prefer to say that science and religion are two independent ways of viewing the world. He rejects that approach, feeling that if biblical literalists wish to do science they are going to have to undertake some challenging mental gymnastics, and he does not feel that it is up to scientists to do the heavy lifting for them.

Needless to say, his article prompted several letters to the paper in response. Six were printed the following week, one by a Physicist, who largely agreed with the points made in the article, though at the end he waffled a bit. The other five letter writers disagreed strongly with the article in different ways, but all agreed that science does not provide a complete explanation for life, so some sort of religious explanation is needed. But none of the five seemed to understand (as the Physicist mentioned) that religion does not conform to any basic natural laws, so it does not tell us anything useful about existence. If you object to the word "useful," remember Marcia McNutt's statement at the beginning of this book. We are interested in ideas that conform to the laws of nature because those laws tell us how the world works.

Bertrand Russell once wrote that if he claimed that between the Earth and Venus there is a teacup orbiting the Sun, there would be no way to disprove the belief, even though there would be no way to verify it. From his claim, he said he could construct a consistent set of religious beliefs, which would have as much going for them as those preached by contemporary religious leaders.

If that sounds silly, try to come up with a way to falsify the beliefs of any set of religious beliefs. Religion is not equivalent to science. Anything that is not based on conforming to the laws of nature should not be considered as an appropriate component of any science course in the public school systems.

In demanding that their views about Biology be taught in public school science classes, Creationists claim they want "equal time" to counter the material in the curriculum on Evolution. But they have nothing to present that does not consist of pure, unadulterated religious beliefs. They claim that what they would present would be scientific, but they have no support for any of their claims from any scientific discipline. Nothing.

When presented with that statement, they fall back on the claim that evolution cannot be "proved," so it does not represent science; it represents a religion, and they just want equal time for their religion. Aside from the fact that they are implicitly saying that what they want to present involves religion, which cannot be sanctioned by any government agency (because of that pesky First Amendment), they are admitting that nothing in the bible can be proved. I have to wonder why they believe anything in the bible if nothing in it can be proved. But more on that later.

Going on, Creationists insist that the Earth is only a few thousand years old. Some say 6,000 years and others are willing to admit that it might be as much as 10,000 years. The variation seems to depend on how carefully the ages of the patriarchs are added together. Of course, these numbers depend on something they assume, which is that the narratives in the Old Testament are literally true. So, forget about how many years they come up with. The important thing is that they are assuming that the narratives are a reliable source of history. One would think that if they are so sure about that assumption, they would be able to provide some evidence for it for the rest of us to consider.

After all, none of us were around at the time to witness what happened in the past, and the Old Testament was written down long after the events described would have had to have happened. So, what is the source of their certainty that the Bible is a reliable history book?

The only answer to that question that I have come across is not much of an answer. They claim that the Bible is God's word, so it must be true. But how do we know it really is God's word? Because, they respond, the Bible says it is. After all, God spoke directly to Abraham and Moses, etc. It says so right there in the Bible. That is a brilliant response, because divine revelation cannot be disproved. Of course, there is no way to prove it either, but that does not seem to matter to them.

The circular reasoning in the last paragraph must be obvious to some Creationists, but that knowledge does not change their attitude, because logic does not trump certainty. The biblical narratives provide support for the kind of world they wish to live in, so they do not need evidence to support any of the details.

The British writer G.K. Chesterton once wrote that the value of the fairy tales that are told to children is not to convince them that monsters exist. Their value really is to convince children that monsters can be killed. What Creationists will not accept is that the point of the stories in the Bible is not to provide historical accuracy. Their value is to use stories to develop a narrative that provides a sense of morality. But that seems to be too subtle for them.

I feel sorry for Creationists, because their need to defend a literal interpretation of the Book of Genesis forces them to reject modern science. It is sad when intelligent people find it necessary to defend something that is erroneous, just because the errors involve things they have been told were true since they were children. Their rejection of Evolution is a good example.

The explanation for how evolutionary changes have resulted in so many different kinds of living organisms on the planet makes so much sense that it is hard to believe that anyone would reject it because they prefer some Bronze Age myths that have no evidence to support them. But when faced with scientific results which contradict what they have grown up hearing and believing, Creationists reject the science. An excellent example is their rejection of genetic changes that are transmitted over time, resulting in new kinds of organisms - what we call the evolution of species by natural selection. The mechanism Darwin proposed for how evolution works.

Creationists insist that there is no proof that the diversity of living organisms we see around us is due to evolution, so it should not be taught in public school science classes. They say that minor variations obtained in studies of fruit flies (what they call "microevolution") are not the same as the development of entirely new species of mammals such as humans (what they call "macroevolution"), and no one has ever seen a new mammalian species develop.

To maintain their beliefs, they ignore the two facts that are the basis for evolution by natural selection: genetic changes and very long time periods. Genetic changes occur because mistakes are sometimes made as DNA is copied when an organism reproduces. That is a fact. It is true that the cells in our bodies have ways to correct those mistakes, but they are not always successful. The mistakes that are not corrected result in changes in genes, which are segments of DNA that provide instructions for the cells on how to do things (such as how to make a specific protein). Sometimes those changes are benign, and sometimes they are fatal to the organism's offspring.

Long times are provided by the 4.5 billion-year age of the planet. All it takes is a copying mistake that gives one individual a mutation that is advantageous now or which becomes advantageous sometime in the future as the environment changes. As long as the mutation is benign it can be carried over for many generations until for one reason or another, it allows the organism to be more competitive than others in its group. In time-spans much longer than the lifetime of the organism, the descendants (each of whom carries that benign mutation) can acquire many new mutations, so in a sense, over what Paleontologists call "deep time," lots of experiments occur because of the copying mistakes. If you keep tampering with something, as long as you don't kill it, you should not be surprised if the end result does not resemble the original version. So, there should be no reason to expect the organisms surviving at the end of very long sequences to be "compatible" with those at the beginning.

The point is that Evolution is based on a mechanism that has been observed (copying mistakes during reproduction), and on time periods sufficiently long to allow the mutations to eventually be matched with an environmental change that occurs. Two ideas that make sense.

Instead of challenging scientists to prove that new species have developed by this mechanism, Creationists should be ready to prove their own suggestion for the diversity of living organisms on the planet, namely, intervention by a supernatural being as the cause of what we see around us.

Go on folks, prove it. Without relying on the Bible.

Evolution is not the only scientific subject Creationists dispute, but it is the most important to them because it pertains to the origin of humans. However, because they cannot "prove" that humans did not "evolve" without citing the Bible (and arguments based on the Bible are not given any credence in courtrooms), they pick away at other aspects of science in order to convince people that evolution could not possibly have happened. Not that it *did not* happen, but that it *could not* have happened. Questioning studies that indicate that the Earth is very old is their favorite strategy. That is a sound rhetorical ploy, but it does not work.

So let's start with the fundamental position of the Creationists, which is that the Earth could not possibly be billions of years old because an "old Earth" contradicts what the Bible seems to say, and biblical inerrancy is the only foundation many people have for their beliefs.

There is a problem with biblical inerrancy. To say that the Bible is the Word of God so it must be literally true, and then say that we know what God has done and what He wants because it is found in the Bible, involves circular logic. So let's avoid interpretational quagmires, and look at some rather mundane facts.

We'll start with some trees. The age of a tree is determined by drilling a hollow tube into it, extracting the tube, and counting the growth rings. What could be simpler? Using this technique, some bristlecone pine trees have been "dated" at 9,000 years, and the age of a creosote bush in California has been determined to be (again, by counting rings) 11,700 years. It is hard to deny that an accurate age of a tree can be obtained by counting tree rings. No assumptions are needed - just arithmetic.

Now here is the problem. Those who add up the ages of the people in the Bible claim that the Earth is only about 6,000 years old. But if the Earth is only 6,000 years old, that creosote bush is twice as old as the planet. How can that be? Where did all those extra annual rings come from? When confronted by numbers such as these, some biblical literalists hem and haw and say that the planet could be "about" 10,000 years old. But that adjustment is not good enough, because it ignores what had to have happened during the biblical "Flood."

But first, some more facts. Trees are very hardy plants. Anything that can survive for thousands of years must be hardy. But one thing they cannot deal with is to be submerged in water. When a dam is built to flood a valley to create a reservoir, all of the plants (including trees) die when they are submerged. They die because they don't get the sunlight they need to "power" their metabolisms using photosynthesis, and they are not designed to extract the carbon dioxide they need from water instead of air. No terrestrial plants can survive being submerged for extended periods of time.

So, if the planet was completely covered by flood waters that were deep enough to cover the highest mountains, why didn't the ancient trees die when submerged? I don't recall reading in the book of Genesis that everything on the planet died except the passengers in the Ark and the bristlecone pines and creosote bushes. Or that some dispensation was granted to plants so they could survive being submerged.

The existence of those trees should convince anyone who believes in facts that a world-wide flood did not occur. And if the story of the flood was made up, why should anyone believe in the chronologies that give an age for the planet of about 10,000 years?

Note that I have not made any assumptions. I just presented some facts which are hard to deny. By structuring the discussion in this way, I am asking biblical literalists to reconcile something they believe strongly (the inerrancy of the Bible) with something that cannot be denied (that terrestrial plants cannot function underwater). This process should induce a great deal of cognitive dissonance, which is good because it forces one to think.

It seems that in order to invoke the Bible as evidence for a "young Earth," one has to ignore things that cannot be denied. How much confidence can we have in a person's rationality when they do that?

It makes sense to say that the subliminal learning that occurs in one's home environment might explain why some people always vote for one political party and other people for another; why some people are Red Sox fans, while others root for the Yankees; why some people are willing to interpret the Bible literally, while others are content to accept its moral teachings without insisting that the cosmology underlying it is accurate.

But that explanation does not explain why the majority of people who are called Creationists, that is, Christians who interpret the Old Testament literally, are concentrated in America. The segment of the population these people represent is nearly zero in other developed countries, such as Canada, Great Britain, France, Germany, the Scandinavian countries, or Australia, whereas in America, Creationists are abundant. Why just us?

Surely, unstructured, subliminal education occurs in homes in other developed countries, and there are Christians in all of them, but if Creationists exist there, they don't share the fervor of those in America, so they are well-hidden. So what is the reason the movement is so strong here?

What is it about Evangelical Christianity in America that provides a home for people who can embrace 21st century technology (smart phones, flat-screen televisions, navigation technology in their cars), but selectively reject aspects of Physics, Geology and Biology (radioactive dating, evolution) that are not consistent with stories first written down for a tribe of Bedouin in the seventh century BCE?

Why do Christians in other countries not peck away at science in the way that the American Creationists do? There is something here that I am missing.

Recapitulation

The material on the last eight pages deals with a group of people whose religious beliefs prevent them from accepting the results of some scientific studies. Some studies, not all. They peck away at any branch of science that implies that their literal interpretation of the stories told in the Old Testament is not justified. In effect, they tell scientists that "you cannot prove "this" or "that" to my satisfaction, so nothing you say about the age of the earth or the existence of evolutionary changes can be true." They refuse to accept the idea that unconditional "proof" is never available, but that provisional knowledge based on existing understanding of natural processes is sufficient to reject their beliefs that myths from the Bronze Age are historically accurate.

People who are willing to accept the advice of stock market analysts about investing their assets refuse to accept the conclusions reached by scientists based on innumerable carefully conducted experiments.

They reject the concept that if it is not possible to devise a test that might show that a belief is false, the belief is suspect because it will not be possible to devise a test to verify it. So they cling to beliefs that have no supporting evidence. The only way they can justify their beliefs involves circular logic, but that is sufficient for them.

They reject common sense conclusions based on easily made observations because the conclusions contradict the narrative on which their lives are based. They reject any attempt to convince them that the foundation of their beliefs consists of stories they have been told since they were children, stories that have no more substance than the stories about mythical events such as the Trojan War.

They reject the idea that the point of the stories in the Old Testament involves establishing moral precepts by which people should live their lives, and that it is not necessary to believe that the events describe in the stories actually happened. They do not see the parallel with Aesop's Fables.

In particular, they reject the idea that evolutionary processes are responsible for the vast diversity of living organisms because they insist that all living things were created by God in the six days of creation described in the Book of Genesis.

Moving Right Along

Creationists don't just say that Evolution did not happen. They contend that it could not have happened, because they accept the biblical account of creation in six days about 6,000 years ago. Now let's look at some of the ways to argue that Evolution makes sense.

I did not take any Biology courses in college, but for several years I have been reading about different aspects of the subject. And because my background is in Geology, I am interested in the subject of the evolution of life and how genetic studies illustrate how living organisms evolved over the last few billion years.

The genetic "code" is the language used by each cell in a living organism to interpret the information in a DNA molecule to make the proteins a living cell needs to function. This code is universal in the sense that most if not all living organisms use the same code. The fact that everything from E. coli bacteria to humans use the same "program" to translate the messages in their DNA molecules is hard to explain unless all living things have a common ancestry.

And, the common ancestry of all life provides a strong argument for the validity of the Theory of Evolution. If every living organism uses the same code (computer program, dictionary, whatever we call it) to function, it must be because all of us (from single-celled bacteria to humans) originated from the same kind of organism (a single-celled bacterium).

Over the eons, more complex life forms evolved as mutations caused rearrangements in the sequence of the instructions, but the "language" did not change. Think of a change due to a mutation in terms of causing a word in a book to be misspelled. Sometimes the change due to a mutation makes no sense and the organism cannot make a protein it needs, so it dies and its group may become extinct. But sometimes the changed word provides a new way to do something, and the organism "finds a new job" (evolves into something new). The point is, that rearranging letters can affect the meaning of a word or sentence without affecting the grammar, so even if the meaning of a word may change, sometimes the sentence continues to make sense. In the language of linguists, a change in semantics does not necessarily change the syntax.

Because mutations tend to occur randomly, there is no way to predict how an organism's group will change over time, or whether its group will even survive. Thus, there is no "direction" to evolutionary changes. And no purpose. All of the species that have become extinct over the eons indicate that.

To repeat, the fact that so many different kinds of organisms use the same molecule (the double-stranded nucleic acid, DNA) to determine how the organisms will function suggests a common origin. And the fact that regardless of which organism we are considering, segments (that we call genes) of that molecule contain instructions "written" in the same language, also suggests a common origin.

And a common origin which leads to an amazing diversity of organisms is evidence that Evolution is more than a hypothesis to explain the biologic patterns we observe today: it is a fact.

If you are interested in a "book" that provides a description of the history of life on Earth, forget about the Old Testament. The DNA molecule in every cell in your body has what you want. It provides a history of life on the planet.

Some examples of that history can be found in a fascinating book by Gregory Chaitin, a Computer Scientist/Mathematician, about Evolution. His background gives him an interesting perspective on this Biological topic. The bulk of his discussion pertains to the human genome, that is, the complete human DNA molecule. Think of the genome as the "Operating Instructions" for humans because it provides instructions for every cell in the body to build the proteins they need to function.

Now, about his perspective. Because he is a computer scientist, he points out that the human DNA molecule looks like a giant software program that has been modified from time to time by different people who had different goals and different programming styles. That is how large software systems look. When changes are called for, it is too time-consuming to start from scratch and write a new version at the assembly-language level, so people cobble together components in high-level languages and do their best to make sure that the new stuff does not compromise what others had done in the past. Very few large software systems are "elegant."

Then he points out that the people who have sequenced the human genome (that is, worked out the sequence of chemicals along our double-helix molecule) found that it contains components (call them subroutines) from sponges, from amphibians, and from fish (which is the reason that at one stage in its development, a human embryo has gills). Later, some jerry-rigging was done to add components for reptiles and later for mammals. Lots of tinkering was involved, but little if anything was jettisoned in the process (hence the gills). It is far from elegant.

Do you see the point? Chaitin points out that the contents of the human genome represent a history of life on Earth (fish, reptiles, mammals). That provides a very good argument for Darwinian Evolution.

Oh, one more thing. About "elegance." Have you ever heard of the "Intelligent Design" argument the Creationists use? The one that goes that the human body is designed so well that it could not have developed by random mutations? Well, a little thought about the human genome should convince anyone that all that is nonsense. All the tinkering involved with developing what became our genome suggests that those who use the "Intelligent Design" argument have never looked at life at the molecular level. The "Operating Instructions" for our species are too kludgy to have been designed on purpose.

In effect, the genome represents a book that has gone through several drafts, but has never been worked over by an editor. Hence, the "kludgy" nature of its text.

When new products are developed, the first ones off the assembly line often have some defects that quality control measures detect, and changes in the production process are made to correct those defects. But biological evolution does not work that way.

William Calvin is a neurophysiologist who has written several books about the evolution of humans with an emphasis on the development of our brains. In one of the books, he discussed the topic from the standpoint of the common logical fallacy "post hoc, ergo propter hoc," which means, "after this, therefore because of this." That is, when B follows A, we assume that B was caused by A. (Note: he did not actually mention the fallacy, but that is the logic he used).

What has this fallacy to do with evolution? Well, when we see diagrams showing the kinds of primates that have lived in the past several million years, arranged sequentially, we see what appear to be relatively primitive creatures, with less primitive creatures above them in the diagram, with human-like ones above them, with us above them. So we tend to think that there has been a steady progression from the bottom of the diagram to us at the top.

It is easy to make the assumption that the progression shown represents a process involving steady "improvements," but that is because we tend to ignore the fact that there are many gaps between the species shown in the diagrams. We assume a direct lineage and that each is "better" than the preceding ones. That implies that some sort of "direction" was involved in the evolution of our species. We were pre-destined. That is the fallacy. Just because we think we are better than what has preceded us, we think that the changes in the sequence were intended to make us what we are.

There is an old adage that says "the first of its kind " (of nearly anything), tends to be a bit rough around the edges. It takes more than one effort to get the thing right. Keep that in mind when you think about our species and our intellectual abilities. In fact, our intellect is a big step up from those of the other primates, but it appeared only recently (about one hundred thousand years ago, during the last inter-glacial period, which is basically yesterday), long after the human brain stopped enlarging. It is highly likely that it is not well-tested yet, and is still prone to malfunctions.

The next sentence is very important.

Our intellect is not well-tested because biological evolution does not perfect things. Got that? Don't expect perfection from evolution. The mutations that produce new things (and the genetic permutations that occur when we inherit chromosomes from both of our parents), are random and unpredictable, so they just produce things with different sets of "bugs," not finished products. We have just recently evolved from the primates that were our ancestors, so we are trying to run twenty-first century software on hardware that is at least one hundred thousand-years old. Imagine trying to run a modern word processor on an Apple II™.

Biological evolution cannot be depended on to get the bugs out of the genetic code (remember, there is no preordained "direction" to evolutionary changes), so we have to depend on cultural innovations to correct defects. But, when we think about the resistance to taking the steps needed reduce future global warming, it should be obvious that cultural innovations are not accomplishing much "progress." Think about the following example.

The first members of the genus "Homo" (Homo habilis) evolved a couple of million years ago. They made stone tools that presumably helped them hunt, so they must have had some sort of culture, even though their brains were much smaller than ours. The interesting point is that their tools did not change for nearly a million years, implying that their culture did not evolve, possibly because their brains did not. Now here we are, with much larger brains, which are a relatively new feature (only about 100,000 years old). Due to the kinds of things we are doing to the planet, our environment is changing more rapidly than our brains are evolving. That is not a good sign.

Analogies are interesting mental constructions. In a posthumous collection of some of his essays, the late Douglas Adams (the guy who wrote *The Hitchhiker's Guide to the Galaxy*) wrote about an analogy between computation and evolution.

Based on his fascination with computers (he was writing in the year 2000) he noted that "complex results arise from simple causes, iterated many times over." By that, he meant that if you plug numbers into a formula, evaluate the result, and then plug those that come out back into the formula, and if you do this for hundreds or thousands of times (hence the word iteration), even what appears to be a simple process can give very complicated results.

Quite often, the pattern that results suggests a random process is at work, not the simple straightforward one you started with. The cause seems to be due to what is called "roundoff error." Each time through the equation the number of decimal places in the value of the solution is truncated a bit. That is, the solution may be accurate to 12 decimal places, but the program solving the equation does not accept inputs greater than 7 decimal places. So what is plugged into the equation again for the next iteration is not quite what was just calculated, and that difference may be enough to eventually tweak the system into giving different results. This was first noticed in studies of fluid motions, and eventually it led to what is now called Chaos Theory.

Here is the analogy. Adams went on to note that those who study the evolution of life are looking at the same kind of process. When an animal is born, its genome is not quite identical to that of its parents because random copying errors occasionally occur. So the animal's future may turn out to be different from that of its parents. There is no way to predict how the change will affect the animal, but if you do this for a thousand generations, you may end up with a transition from Australopithecus afarensis to Homo habilis.

Adams did not take this thread any further, but it is important to note that analogies are not examples of something. That is, they do not represent a "thing." Analogies describe similar relationships. They tell you that the way one thing works is somewhat similar to the way something else works. Somewhat similar, but not exactly similar.

Now let's take this a step further. One of the characteristics of random processes, is that a random sequence does not look random. In fact, if you run one long enough you will see some repetition. That is another example of how studies of iterated sequences can tell us something about the process we call Evolution. From time to time, the genetic changes in different kinds of organisms may produce similar results (analogous to a couple of strings of similar numbers in a random sequence). A good example is the fact that an organ that is sensitive to light (the organ we call an eye) has developed independently seven or eight times in the last 600 million years, in different groups of organisms. These organs range from ones that can just distinguish between light and dark, to the most complex eye we know about - the compound eyes of a fly. Iterating for sufficiently long periods of time sometimes should be expected to produce new features, and sometimes familiar features. As I said, a random process does not look random.

The example of eyes is useful because it provides evidence that evolution is not "directed" in any way. If things always go from simple to complex, from crude to better to best, why would a lowly fly have eyes much more efficient than ours?

An interesting question. If you believe in teleological processes. you should try to answer it.

Evolution by natural selection is based on genetic changes due to mutations that occur as an organism reproduces. Errors sometimes occur as the DNA molecule is copied, and although there are processes in place to correct those errors, they do not always "catch" every one of them. If the changes due to the mutations are serious the offspring will probably die. If the changes are benign, it will survive and the changes will be passed on to its offspring. Over many generations, benign mutations in the offspring of the offspring, of the offspring (etc., etc., etc.) will result in organisms that may differ markedly from the original parents. When the difference prevents them from reproducing with the members without the changes, we say that a new species has formed.

Intelligent Design advocates contend that humans are too complex to have developed by random evolutionary processes. Some sort of purposive design would be needed, they claim, to produce humans. They do not seem to be aware of the genetic similarities between humans and numerous other organisms. For example, when the genome of a human is compared with that of a fruit fly (Drosophila melanogaster), there is about a 60% correspondence, and 75% of human disease genes are also found in those flies. The correspondence is the reason that the flies are used as genetic models in some studies of human diseases such as Parkinson's and Alzheimer's.

So, how unique are humans? One wonders if the "intelligent designer" used us as a model for the flies or the flies as a model for us.

In addition to Creationists who reject the theory of Evolution because it is not consistent with the biblical narrative involving special creation, there are people who believe too much about Evolution. They see the diagrams in books implying a transition from (what we consider to be primitive) ape-like animals to the first members of the genus Homo, to more members of the genus, each one up the diagram appearing more like we appear, to the first members of our species, Homo sapiens. What they do not realize is that the appearance of this transition represents an assumption. There is no way to determine how "smooth" the transition is, so we have no way to determine if we are the direct descendants of any of the particular species shown in the diagrams. The transitions are not smooth because of how fossils occur.

Darwin devoted a chapter in his *Origin of Species* to fossilization. The likelihood that the remains of an animal that dies will be buried and fossilized before it is eaten and its bones scattered is quite small. And beyond that, if it's remains are buried and fossilized, the likelihood that the remains will be uncovered and discovered eons in the future are even smaller. So the data set is relatively sparse, and there is no way to establish definite links between the organisms that have been discovered at different "levels" thus far. Anatomical appearance and the quality of the stone tools they made (if they made any) are the criteria used. In a sense, the objects in the data set are equivalent to isolated "events," but there is no way to establish which ones actually descended from which ones.

So, anyone who notes that "we" are at the top of the diagram, and from that infers that the entire sequence was directed toward producing "us," is engaging in an act of wishful thinking. It appears that we (humans) represent just one manifestation of a massive effort at blind experimentation, which we assume was aimed at producing us. But there is no evidence that we were the goal of the experiments. It is just egotistical thinking.

Before any kind of supposition can be accepted as true, it should pass some tests. For example, before we agree to believe that some event occurred in the past, we should find the answers to three questions:

1. What is the evidence supporting the existence of the event?
2. How credible is the evidence?
3. Is the evidence relevant?

Without answers to these questions, we have no reason to accept the existence of the event. So, let's apply this principle.

Creationists contend that the Earth was created in six days about 10,000 years ago. Alright, let's use the three questions to test their contention.

1. What is the evidence? The only evidence they have is the Book of Genesis in the Old Testament, which is a collection of stories written about 700 BCE based on oral tradition that goes back two-thousand years further.

2. How credible is the Mid-Eastern oral tradition? No more credible than any other collection of folk tales. And considering the fact that most "creation stories" from different cultures differ in detail, none of them can be considered any more credible than the others.

3. Is the evidence relevant? If the evidence is not credible, if there is no reason to believe that it describes what actually happened, how can it be relevant?

There is no evidence that the stories told in the Bible are literally true, and, based on the structure of the document, plenty of reasons to doubt it. These reasons include internal evidence such as inconsistencies within the stories, and a general lack of correspondence between events portrayed in the Old Testament and archeological evidence found in the Near East. But creationists have to believe that the stories are true because to them, the stories provide more than just the foundation of their religion. If true, the stories would demonstrate that all other religions are false, so everyone should subscribe to *their* religious beliefs. That is the basis for their attempts to teach their beliefs in the public schools. They want everyone else to believe what they believe.

So, Creationism has nothing to do with science or history. It is a religious belief hiding behind a facade erected by people who desperately are trying to avoid questioning what they have been told all their lives.

Does change always involve progress? If by progress, you mean that change results in something better, that is questionable. Better than what? Is it always better than what was there before? Could it be better than other options that might have happened, of which we know little or nothing (because they did not happen)? If we don't know what might have been, how can we say that what actually happened is better than what might have happened?

If that line of thought is not confusing enough, now think about long-range changes, over which we have had no way to affect.

A few billion years ago, an important change occurred in a group of organisms called cyanobacteria, which used to be called "blue-green algae" (inappropriately, because they are not algae). But regardless of the name, they developed the ability to use sunlight as a source of energy for converting the carbon in atmospheric carbon dioxide into the chemicals they needed to grow. We call the process photosynthesis, and its waste product is oxygen. The change resulted in what is called the Great Oxidation Event, because for the first time, an appreciable amount of oxygen began to accumulate in the atmosphere. The word "great" denotes large, or "marked," but it is tempting to believe that this change also implies "better," because if it had not occurred, none of us would be here. Nothing that needs oxygen would be here. And the word "event" implies it happened quickly, whereas it actually took quite a while to have any effect. But we have to describe it somehow.

If that change had not occurred, what would be here today? Is there any way to tell? The answer is no, because there is no way to determine how the anaerobic organisms living then would have evolved if different changes had occurred. Nature was "tinkering," which resulted in what we have today. Steven Jay Gould wrote about this kind of tinkering in a book about the fossils of animals found in the Burgess Shale in British Columbia that lived about 500 million years ago. He said that the "body plans" of the fossils included a number which have never been seen since, so a good deal of tinkering seems to have been going on at the time.

There is no way to determine why some "plans" persisted and others did not, and Gould said that if it were possible to "run the tape" again, it would be highly unlikely that the sequence of organisms that persisted would be the ones we are familiar with. Would that have been bad? We might think so, because we would probably not be here today.

Regardless of what we would like to believe, just because humans eventually evolved, does not mean that all of the changes that occurred since the development of photosynthesis in the Pre-Cambrian were intended so that humans could evolve. The results of tinkering are unpredictable.

Lewis Thomas once said that the only thing we can be certain about is that if the tinkering did not occur, we would all be anaerobic bacteria and there would be no music.

So, the words good and bad are irrelevant when thinking about what has happened in the past. Nature's tinkering, produced what we have today. But "good" and "bad" are not irrelevant for future changes, because we have the ability to decide something about the future. For example, we can decide if climate change is going to have moderate or disastrous effects on most organisms living today. There have always been variations of climatic conditions in the past, and some organisms survived while others did not. But the kind of change that is likely to happen if we persist in the profligate use of fossil fuels is a kind of tinkering whose effects can be predicted, and the prediction is that they will be quite severe.

Recapitulation

The material on the last fourteen pages provides support for the belief that the large diversity of living organisms is due to evolutionary changes using the mechanism called Natural Selection, which Darwin proposed in 1859. The evidence, most of which was not known to Darwin, involves the modern branch of Biological Science called Molecular Biology, which involves studies of the DNA molecules (the genomes) of different species.

The sequence of chemicals (called base pairs) on each of the two strands of the DNA molecule is equivalent to a computer program which provides instructions for each cell in a living organism that tell it how to build the molecules it needs and how to function. The genome is divided into segments called genes that are tens of thousands base pairs long that are analogous to subroutines in the overall set of Operating Instructions for each cell. There are about 3 billion base pairs in our DNA, and the genetic code uses "words" that are three "bits" long, so there are about a billion "bytes" or a thousand megabytes in the operating instructions. A cumbersome program to decipher. The instructions are referred to as the genetic code, and all living organisms function using the same code. All of them, from single celled-bacteria to mammals. That suggests that every type of living organism has evolved from some primordial single-celled organism.

Another line of evidence for evolutionary changes is the fact that as organisms reproduce, errors occur in building the genomes of the offspring. When those errors are not fatal, the organism is slightly different from its parents, and as this process repeats over many (very many) generations, the cumulative effect of the errors is the development of organisms that have no resemblance to the original parents. New species develop in this way.

There is no way to predict what kinds of errors will occur, so there is no way to specify a "direction" in which the changes will "take a species." Random errors result in random differences over long time periods.

To repeat something said earlier, those who contend that science does not provide enough answers do not understand that scientists share that belief, and that scientists are always looking for ways to expand their understanding, but only by means of verifiable statements. *Religious beliefs are not verifiable*, so they have no place in a science curriculum. The testing procedure described on page 33 allows us to distinguish between unsupported beliefs and verifiable suggestions.

Distinguishing Between Belief Systems

At this point you have read quite a bit about two kinds of belief systems. One is based on the authority of a text, the Old Testament, which is a book whose oldest parts were written down about the seventh century BCE, which consists of stories based partly on oral tradition and partly on written materials (clay tablets with cuneiform symbols impressed in them) from the Akkadian and Sumerian cultures that extended back to the middle of the third millennium. The second belief system involves what one learns from inquiries about observations. The desire to know why things happen as they do drives one to make suggestions and then test them to determine how well they explain the observations. This approach blossomed in the eighteenth century, and has been responsible for the development and progress of scientific and technological thinking since then.

The distinction between the two belief systems is sharp if we confine the discussion to those who depend on the authority of the Old Testament unequivocally. They believe that the Bible is inerrant because it is the word of God, so it cannot contain any errors. They may admit that they do not always understand how some of the things described could have occurred, but they have no doubt that they did. They feel that faith in the inerrancy of the text supersedes any questions about the need for an explanation of some things.

Of course, many "believers" do not find it necessary to be "Biblical Literalists," so they are willing to accept the moral lessons contained in the Bible without requiring faith in the literalness of the stories. These people have no problems accepting the scientific approach to how the world works. They are not the ones who send letters to the editors of newspapers or go to school board meetings to denounce the teaching of Evolution in high school science classes. They are not the ones who contend that the Earth and everything on it was created in six days, as described in the Book of Genesis. That is, they are not the Creationists described above.

Keeping in mind the difference between the people who rely on the Bible solely for moral lessons and those who insist that everything in it is literally true, the distinction made thus far in the book is between the Literalists and those who accept the results of scientific studies to explain how the natural world works and how it evolved from its earliest times to today. Note that the believers who accept scientific results are doing so seamlessly, in that they do not have to erect walls between different parts of their belief system. They can ride in a car or airplane without the need to cherry-pick the parts of science they will accept.

You do not have to jettison religious beliefs to accept the findings of science, and many people do not. But you should reject stories based on Bronze Age myths that cannot have happened either because they are physically impossible (the Sun orbiting the Earth) or because there is no evidence to support them (the Flood). The moral teachings in the holy books of the major religions need not be rejected, so you do not have to reject science to "find the Lord." Just remember that it is important that your beliefs conform to the basic laws of nature.

A QUASI-RELIGION: FREE MARKET ECONOMICS

Now we switch tracks. At least, we switch some of the labels. The rest of the book deals with the belief that we must ignore recommendations based on scientific studies that may threaten the operations of our economy. Some of those recommendations involve steps needed to protect the health of the population, and others pertain to ways to prevent climatic changes that threaten most of the kinds of living organisms on the planet. You may well wonder why the economy is more important than the health of people and of the planet. But there it is.

As before, we are dealing with two contesting sets of beliefs. On the one hand, we have a group of people who accept the results of scientific studies that indicate that pollution due to industrial discharges is a threat to the health of the public, and that carbon emissions from power plants are causing changes in climate that will have a series of consequences that will be detrimental to most living organisms on the planet. The members of this group believe that protecting the public from pollutants and preventing global warming from continuing to induce changes in climate are more important than protecting the "bottom line" of major corporations. On the other hand, another group of people consists of advocates

41

of free-market economics, who reject any suggestions, no matter how sensible and logical they may be, that impose any restrictions on economic activities. In particular, they insist on rejecting suggestions based on scientific studies, even in the face of obvious negative repercussions of untrammeled economic activities such as public health crises caused by pollution and the warming of the planet due to carbon emissions.

So, one belief system is based on inquiry and concerned with protecting the public, and the other is based on traditional beliefs about the economy which are analogous if not equivalent to revealed truths. In effect, free-market economics is a quasi-religion. So let's explore this set of beliefs.

The closest thing to a revealed truth in our society is the importance of avoiding anything that would restrict economic growth. One measure of that growth is the Gross Domestic Product (GDP), which is the market value of everything produced in the country in a given year. One value for everything produced in the country. When the value goes up we are told that the economy is improving, in which case the nation is progressing economically, because clearly the nation is "worth more" when the GDP goes up. It seems to make sense, but when you conflate information about many thousands of different items into one numerical measure, you lose a great deal of resolution, which means that you do not really know much about the economy itself. But regardless of the loss of resolution, it is easy to get used to using the GDP as an indication of how the economy is doing, because changes in its value are obvious, even though it is easy to forget that the value of a single number tells you nothing useful when it represents too many disparate things.

But the real problem with considering the GDP to be an important index is that it is based on Production, and nothing else. If its value increases from year to year, people are happy because they think in term of progress because things are being made, and presumably, sold but they do not know how it is growing, or what negative implications are associated with its growth. The costs associated with growth are not accounted for in the calculation of the GDP's value.

Production implies that raw materials are being extracted from the ground, manipulated in various ways by workers and machines to turn out commodities which can be transported to different locations and sold to the public. Every step in that sequence requires the use of energy, and that invariably involves the discharge into the environment of waste materials. Emissions of gases to the atmosphere, discharge of chemicals into water bodies, and burial of solid wastes into landfills (that eventually leak) are all involved at the tail end of the production process, but unfortunately, they are not usually considered in the cost/benefit calculations that determine whether the production of goods will be profitable.

Emissions and discharges into the atmosphere and water bodies result in pollution which imposes costs on the public, which has to pay to deal with them. The regulations that reduce, if not prevent, that pollution saves the nation from the need to pay those costs and even more important, reduces, if not prevents, health problems associated with breathing foul air and drinking polluted water.

Consider this example. If repeated exposure to foul air causes me to develop respiratory problems, I will see a doctor for treatment, which might include medications, and the cost of those medications will count toward the value of the GDP. So the emissions that caused my respiratory problems have indirectly contributed to a rise in the GDP. Does it make sense to say that my declining health contributes to an "improved" GDP?

Now consider this. A federal regulation that prohibits the emissions that caused my respiratory problem would be considered "burdensome" to the industry responsible for the emissions, even though my health would improve. It seems that industries have a financial incentive to pollute the atmosphere and subject the public to conditions detrimental to their health, so naturally, they lobby Congress to do away with restrictions on their activities. But there is no way to put my improved health into the tabulation. Any emphasis on growth that ignores the true costs of the growth distorts the economic measure we call the GDP. And regulations that reduce or prevent health problems among the population should be considered to be a form of protection, not a restriction.

What follows represents some detailed comments on the distortions caused by biased thinking that I posted on my blog over the last year or two. As before, there will be some repetition, but not too much.

It seems safe to say that America's national belief system is based on Capitalism, not on the Judeo-Christian tradition, or on the scientific thinking developed since the Enlightenment. "The gospel of prosperity" pertains to more than some Protestant sects, because the belief that we can all be rich and powerful as long as the government does not get in our way permeates society. This is the way of thinking promoted by corporate America, because it relieves the executives of corporations of feeling any responsibility for our society's inequitable distribution of assets and incomes caused by rapacious business practices. We are told that everyone has the same opportunities, so we can all succeed if we take advantage of them.

This set of beliefs is analogous to the "prosperity gospel" ideology that is based on the principle that riches will be yours if you just accept Jesus as your savior. In the meantime, just send some money to the ones who preach this doctrine, in order to qualify for the largesse that Jesus has waiting for you. Anyone can "succeed" as long as they meet the qualifications. In the case of the economy, "taking Jesus into your heart" translates into the belief that we can be successful economically if we are smart, talented, and hard-working.

Economists and politicians tell us that it is important that the economy be allowed to function as efficiently as possible, so everyone will have the potential to achieve prosperity. That is why they say that the government should not impose burdensome regulations on corporate activities, because they tend to stifle entrepreneurship and the concomitant development of jobs and economic prosperity.

We are told that the regulations which increase the cost of doing business, or ones that prevent companies from exploiting opportunities to make a profit, are not appropriate, because when government is *not* that heavy-handed, the economy will grow and everyone will be able to take advantage of the myriad opportunities that will develop – at least, those of us who are talented and work hard.

These concepts underlie the anti-regulation "sermons" that lobbyists for (to mention just one organization) the U.S. Chamber of Commerce preach to Congress incessantly. These lobbyists are the secular version of preachers at tent-show revival meetings, and although they do not manage to heal the lame and cure the sick, they do manage to convince enough politicians that their doctrine is sound, so it has prevailed since at least the Reagan administration. Of course, the fact that they spread campaign contributions around liberally, does not detract from their message.

One aspect of the worship of economic growth involves the mechanism by which the GDP can grow, which is the "free-market." Those who believe that the free-market can deal with all problems involving the economy would have us believe that individuals acting in their own interests, will always make decisions that are the most favorable to them. Extending that thought to an entire population, the idea is that when everyone acts freely in their own interests, the overall result is the best for the entire society.

The basic assumption is that all of these transactions are independent of all of the others, in the sense that what I buy does not affect your choice, and that the prices are set by the merchants independently of each other, and therefore the prices can be negotiated, so if you do not want to pay the asking price you can bargain with the merchant, knowing that if you continue to walk down the street, you may well find someone offering the same item for a lower price. Of course, the merchant knows that too, so he will be willing to bargain, and everyone will be satisfied with the result.

This ideal is basically what may go on in the mid-east, say in an Arab market in Algeria, but it happens in developed countries only in flea markets. What you usually buy, especially in a "big- box" store, has a price tag on it and that price is fixed. No negotiations are possible. In addition, the price for that item is probably essentially the same at any store you enter, because the production and distribution systems are the same for nearly any vendor, so the fixed costs are about the same, and prices will not vary too much.

So, we don't have the classical "markets" in our society. But if we think in terms of, say, 100 million consumers, even the big-box stores can be thought of as individual vendors, who keep track of what sells and what does not, and at what price. That information determines what will be reordered and will affect the price for an item in the next shipment. So, in a sense, the random independent acts of numerous consumers simulate what goes on in a traditional market.

Now consider why governments tend to impose regulations on business activities. Children tend to chew on toys, so when you buy a toy that was imported, you should not have to worry whether the paint used on it contained lead. Regulations that prohibit the sale of dangerous items protect your children, so even though they may raise the cost of the item, the amount is a miniscule amount for any particular item bought, but the cost to you if your child is poisoned by lead paint is considerable.

That is why we need regulations. If you think of them as "protections," you understand their need, even if the executives at a corporation grumble about them being "burdensome" because they raise the price of their products, which may reduce the number sold. Regulations that interfere with "market activities" are not intended to be burdensome; they are intended to protect people, even if those people may not be aware of the need for that protection.

When lobbyists for large corporations try to convince legislators that regulations will hurt the economy, what they really mean is that their employers do not want anything to interfere with their profit margins, regardless of any long-term health effects suffered by the public.

Another reason we need regulations is because politicians pass laws that that require implementation.

To explain that statement, it is useful to distinguish between laws and regulations. A law such as the Clean Water Act is intended to set standards for what can be discharged into the nation's waterways. But merely specifying that discharges should not detract from water quality is not enough: identifying what cannot be discharged and specifying the maximum concentration of anything which can be discharged is also necessary. Of course, the ability to determine those things is well beyond the capability of members of Congress, or the staff members of the committees that write the laws.

For that reason, laws dealing with environmental quality assign the details to agencies in the Executive branch, such as the Environmental Protection Agency (E.P.A.). Experts in the necessary disciplines such as water quality, atmospheric science, etc. in that agency review the relevant scientific literature to determine at what concentrations, different compounds can be dangerous. They also provide grants to university-based scientists to do tests to determine the toxicity of different compounds. With all this information, they set the appropriate "safe" levels. These levels which are published in the Federal Register are part of proposals the agency feels are appropriate ways to prevent pollution. Once comments about them are received and dealt with, they become regulations and rules which are treated as binding, because they are called for in the legislation passed by Congress.

The environmental laws passed by Congress also assign to the Executive agencies the task of monitoring compliance with the regulations, so regional offices are set up that are staffed by experts who interact with their counterparts in state agencies. In principle, the regulations developed by the Executive agencies represent the most current scientific findings of the relevant professions. Actually, the steps needed to develop the regulations and get them approved can take several years, but there is no way to reduce that to several months, because of the requirement that sufficient time be allowed for comments from the industries involved and possible lawsuits from some industries. But they are as current as is feasible.

The point is that the regulations issued by the agencies in the Executive Branch are not arbitrary. They are not examples of the Executive branch pre-empting the role of Congress. Their purpose is not to stifle economic growth. Their purpose is to respond to problems specified by laws passed by Congress, and their development takes into account the best scientific knowledge available and also objections from the industries involved.

To complain about "burdensome" regulations on economic grounds is to ignore the problems the regulations address and the science behind their development. So, the emphasis in society on free-market activities that are supposed to promote economic growth, often conflicts with rules and regulations established to protect the public from some problem. These rules may involve water and air pollution, workplace safety, the thickness of the pavement on interstate highways, the dimensions of exit ramps on highways, etc. The list could go on and on. They are responses to problems that are not trivial, regardless of how "burdensome" they may seem. And when the agencies are staffed and headed by professionals, their activities can be labelled as "overreach" only by people whose oxen are gored.

So to summarize, the justification for many government regulations is based on the results of scientific studies. Remember Marcia McNutt's comment in the Introduction about science. Science provides a way to decide whether we should believe something. If there is a reason to believe that some business activities threaten the health of the public, it makes sense to restrict those activities.

Remember the example of toys painted with lead-based paint.

It is one thing to say that clean air regulations protect the public, but it is another to explain the details of why they are needed. The next few paragraphs give one example of the science behind the regulations.

In large cities, household trash that is picked up by the local sanitation department is taken to an incinerator, where it is burned rather than dumped in a landfill. Burning is felt to be a better choice because the less that is dumped in the ground, the longer the lifetime of the landfills. But burning puts waste products into the atmosphere, so it is important to prevent air pollution from the incinerators.

To explain what goes on in an incinerator, I will use the example of a pesticide that is no longer sold to the general public, but which is easy to describe. The DDT molecule looks like a dumbbell – a thin flat region with two circular blobs stuck on the ends. The middle region consists of an ethane molecule which has two carbon atoms with six hydrogen atoms stuck to them. But three of the hydrogens have been replaced with chlorine atoms (to make it a poison). The circular blobs at each end of the central zone replace two of the other hydrogen atoms normally present. Each of the "blobs" has eight carbons arranged in a ring with hydrogen atoms stuck to them. But as before, some of the hydrogens (two on each "blob") have been replaced with chlorine atoms. The "blobs" are called Phenyl groups.

So DDT has two phenyls which are attached to the ends of an ethane molecule. Each of the phenyls has two chlorines stuck onto it, and the ethane molecule connecting the phenyls consists of two carbons to which three chlorines are attached. Hence the name:

Dichloro,Diphenyl,TrichloroEthane. Two chlorines on two phenyls, stuck on each end of an ethane molecule that contains three chlorines. DiDiTri. DDT.

Now, the shape of the DDT molecule is somewhat similar to a number of organic molecules to which we are often exposed in one way or another. So, the cells in your body, say skin cells, will recognize the thing if you come into contact with it, and the molecule will be absorbed and end up in your bloodstream. Of course, you don't want the stuff in your system, even at low concentrations. After all, it is a poison. So, we need to prevent you from coming into contact with it.

If the temperature of an incinerator is sufficiently high, the DDT molecule will be broken up into individual carbon, hydrogen, and chlorine ions (the difference between ions and atoms is not important here), and will float out the top of the chimney into the air and be carried away. As individuals, they are harmless.

But, if the trash in the truck is wet when it goes into the incinerator, the temperature in the "burner" drops, and may not be high enough to completely break up the DDT molecules. So, because the combustion is not complete, fragments of the molecules end up going out the chimney. The fragments are referred to as "Incomplete Combustion Products." This happens with every complex molecule dumped into the incinerator, so no one really knows what the mix of these "products" consists of.

It is expensive to maintain the temperature in the burner at a very high level, so without regulations requiring high temperatures, people downwind from municipal incinerators would be at risk of developing serious respiratory problems. Even with the regulations, what comes out is not always completely "disassembled."

Of course, municipal incinerators are not the only source of emissions to the atmosphere. Nearly any large industrial facility releases various kinds of compounds from its chimneys to the atmosphere. Preventing these emissions is expensive, usually involving passing the emissions through what is called a "scrubber," which is not just a filter: it is a very elaborate and expensive industrial processor, which is why industries complain bitterly about the requirements imposed by clean air regulations. The purpose of the scrubbers is partly to prevent some things from being discharged and partly to keep the levels of what *is* discharged low enough that the chances of serious effects from air pollution are minimized.

When politicians talk about doing away with "burdensome" regulations, they are trying to avoid telling you that they care nothing about your health. Keep that in mind any time you hear a politician talking about the need to make industry more competitive by doing away with regulations. Asthma is a burden to people who breathe polluted air.

Now that you know something about why we need clean air regulations, let's consider different aspects of the controversy over them.

The first book specifically devoted to Economics was written by Adam Smith. Its short title is *"The Wealth of Nations,"* and it was published in 1776, when economic transactions were concentrated in small businesses. The factory system that dominated the English economy in the 19th century had not yet developed, so very few of these businesses employed large numbers of adults and children working in sweatshop conditions, and there was no need for the kinds of regulations imposed on industries today. There were no restrictions on the number of hours anyone could work; no minimum age for child labor; no workplace safety rules; no minimum wage laws; no workers' compensation laws; no right to join a union; etc.

Within about fifty years, all that changed. That is, the working conditions changed, because as the factory system developed, working conditions differed markedly from those in small shops.

Tony Judt, in a book titled *"Ill Fares the Land,"* discussed the development of regulations that changed working conditions late in the 19th century, which not only improved working conditions, but because there was much less labor unrest, the changes also improved the profits of the corporations. The experience in England suggests that rather than being "burdensome," regulations can be beneficial to both the workers and to the "bottom-line" of their employers.

Now let's go beyond the workplace and consider regulations that target the effects businesses have on the general population.

When you believe that everything is a commodity, you automatically assign costs to them, and feel that they represent a business opportunity if no one else is using them ("exploiting" may be a more appropriate word than "using"). For example, if you find that some property you own contains some mineral resources, you automatically think that it would be foolish not to extract them, turn them into something that can be sold, and reap the profits of doing so. Many people would agree with that, but I suspect that few would agree with corporate actions that exploit "common" resources.

There are many examples, but for now, let's confine the discussion to the atmosphere. We all use the atmosphere. We need the air to breathe, because unlike plants, we cannot use sunlight to run our metabolic engines, so we use the oxygen in the atmosphere for that purpose. Clearly, the fewer extraneous molecules we inhale along with that oxygen, the better off we are, so it is important to avoid introducing those extraneous molecules (let's call them pollutants) into the atmosphere. We call the atmosphere a "common" resource, because no one "owns" it, and for that reason, no one is entitled to utilize it in ways that prevent others from using it as they always have.

Common resources have been studied by social scientists interested in how different societies have avoided misusing them. The late Eleanor Ostrum, of Indiana University, studied cultures in Switzerland and in Japan to learn how they managed to use but not misuse their limited resources. And she received the Nobel Prize in Economics for her studies. Briefly, she found that regulations stipulated who could use a resource and by how much it could be used. By constraining peoples' tendency to overuse something just because it is available, the cultures Ostrum studied persisted for close to a thousand years. Another example involves Pacific Ocean cultures that achieved the same results – as long as their populations did not grow excessively. The conclusion of these studies was that cultures dealing with a finite resource base persist for long periods of time only if they impose some restrictions on how common resources are used.

The examples Ostrum cited should be considered when we think about our use of the common resource we call the atmosphere. No one owns it, so no one should be allowed to misuse it. That is the reason the E.P.A. develops regulations pertaining to air pollution from industrial sources (using the Clean Air Act), and carbon emissions from power plants (using "clean air" rules developed by the Obama administration and suspended by the Trump administration). It should also be the criterion the Courts should use in evaluating the legal arguments pertaining to the regulations.

The "rights" that corporations contend allow them to do whatever they wish, to avoid monetary costs associated with not releasing pollutants to the atmosphere, are fictitious. No one has the right to destroy a resource that they do not own and one that others need for survival.

Experience shows that regulations are the only way to prevent inappropriate use of common resources.

When the E.P.A. started publishing proposed regulations pertaining to carbon emissions from coal-burning power plants in the Federal Register, the Republicans in state houses and in Congress, prompted by the fossil fuel industry, started screaming about "overreach." That reaction was in spite of the fact that a federal court decision (which the Supreme Court chose to let stand) specified that not only was the E.P.A. *permitted* to regulate carbon emissions, but it was *required* to do so by the Clean Air Act. The lawyers for the fossil fuel and electric power industries filed suits against imposing the regulations based on their cost. Of course, the lawyers were careful to limit the cost considerations to those which would apply to the industries, and not consider the costs to society of breathing polluted air.

Cost-Benefit analyses are required by the National Environmental Protection Act (NEPA) of 1969, but that is not mentioned in the lawsuits, because that law requires that the analyses consider all of the costs associated with a policy; it requires that alternatives be listed and evaluated; and it requires that the intangible costs to society of implementing or not implementing a policy be considered.

Of course, the fossil fuel industry's lawyers do not want to go down that path. They emphasize only the monetary costs to the industry of curtailing emissions, because to include all of the costs would require that they include the moral cost of not considering the effects of carbon emissions on global warming and the resulting climate change. If you wonder how to put dollar values on moral costs, it is not easy, but Social Scientists have developed very elaborate techniques to factor intangible costs into cost-benefit analyses.

The so-called "war on coal" is really a response to a war on society by a group of psychopaths who care nothing about the future their and your grandchildren will face.

If you don't understand why regulations dealing with economic and industrial activities are important, think about the chemicals that are found in the drinking water in some cities. Flint, Michigan is the example everyone should be familiar with, but there are many more examples. Because of the agricultural activities in the Mid-West, herbicides and pesticides are found in streams and rivers throughout the region. Some of which feed the reservoirs that supply drinking water. For example, very small amounts of the herbicide Atrazine, which is used in corn fields in Northern Indiana have been found in the drinking water in Indianapolis. But agricultural chemicals are not the only ones that are found in the nation's waterways, some of which provide drinking water for other cities. Here is a list of some of the other kinds of chemicals that are found in streams and rivers in the U.S.

prescription drugs (from antibiotics to anticonvulsives)
caffeine
acetaminophen
steroid hormones
triclosan (an antibacterial)
Prozac (in fish)
compounds from birth control pills and detergents

These compounds, excreted by users or dumped down drains, manage to slip through municipal wastewater treatment systems, most of which are limited to removing disease-causing organisms, (secondary treatment), though many of those systems are being upgraded to also remove phosphorus and nitrogen (tertiary treatment). The reason to remove phosphorus and nitrogen is that they are nutrients for aquatic plants, so they stimulate algal blooms which use up the oxygen in the water, which sometimes kills other forms of aquatic life. The compounds listed above have been found in waterways, sediments, landfills, and municipal sewage sludge (which is often converted into agricultural fertilizer).

The only law that requires keeping track of dangerous chemicals in the environment is the Toxic Substance Control Act (TOSCA) that was passed in 1976 and which has never been updated. It mandates a registry of industrial compounds that may be toxic, but it does not require testing of any of them before they are used. The last figure I saw indicated that the registry contains 84,000 compounds, only a few of which have been tested for health effects on humans. But thousands of chemicals are produced every year and the E.P.A. has never been given the funds to develop anything more than a cursory testing program.

The Food and Drug Administration (F.D.A.) does not regulate the spread of pharmaceuticals in the environment and the E.P.A., under TOSCA, ignores their presence in waterways because they are not "industrial" chemicals. No one knows what effects even trace amounts of these chemicals have on the health of people or the environment.

One reason that these chemicals slip through the regulatory net is that there are so many federal agencies, each of which has responsibility for a small part of the "picture." Their responsibilities are spelled out quite carefully in the legislation creating each one, but coordination is not normally part of the mandate Congress gives to each of them. So the E.P.A. is responsible for some chemicals, the Department of Agriculture for others, and the F.D.A for others. No agency has a budget adequate to even do the job they are mandated to do, much less take on more responsibilities.

To appreciate how fractured the system is, think about a big city police department. Imagine one squad is responsible only for giving traffic tickets, another is responsible only for responding to accidents, another is responsible only for emergencies, etc. The size of the force required to police the city would be excessive and the possibility would exist that some would stand by and not respond to certain events because it was not part of their responsibility.

That is how we deal with environmental degradation. The science behind environmental studies is straightforward, but the belief that regulations impede economic activities takes precedence, because the only lobbying done to preserve the quality of common resources such as the atmosphere and the water bodies is by organizations whose members are denigrated (often called tree-huggers) by the organizations that have vested interests in utilizing resources, regardless of the consequences of the uses. Needless to say, the sizes of the budgets for lobbying by these two types of organizations are markedly different.

The standard rationale for doing away with regulations is that forcing factories to reduce emissions of pollutants would increase the cost of their products, with no concomitant increase in efficiency, so regulating these industries would be inflationary. In other words, we should be willing to put up with some discomfort as long as someone is making money, even if we ourselves are not, because according to Adam Smith, everyone in society benefits when someone prospers. I wonder how many people accept that premise.

Republican politicians never tire of complaining about how government regulations stifle entrepreneurial activities and job creation. But they never complain about the best known regulations - the Ten Commandments. The Commandments tell us what to do (love thy God) and what not to do (kill, steal, covet thy neighbor's wife). Of course, some politicians don't seem to remember that last one, but that's another story.

The purpose of the Commandments was to bind the Israelites together by providing a structure to their society. It may have been difficult at times to avoid killing your neighbor after he insulted you, but giving in to that impulse would disrupt society. Similarly, stealing something from a neighbor is not conducive to a smoothly functioning community, nor is playing around with a neighbor's wife.

In the same sense, regulations authorized by the Clean Air and Clean Water acts provide a structure to our society which prevents corporations from making our lives unpleasant and unhealthy just so they can make a little more money. Poisoning the air or water is about as disruptive as one can be, and regulations are one way to prevent such things in order to achieve the goal of a smoothly functioning society.

For some reason, the analogy between the Ten Commandments and environmental legislation does not seem to be obvious to those who grumble about "burdensome" regulations.

It is hard to avoid stories in the media concerned with conflicts between the states and the federal government. I keep seeing complaints in the media from state legislators about regulations that are called "unfunded mandates," imposed by Washington that make it harder to attract industry to their state. Other complaints pertain to rules that force power plants to reduce their carbon emissions, rules which are construed to be a "war on coal," because the burning of coal is the major way carbon dioxide is added to the atmosphere. And yet, without national rules, the states have no incentive to avoid polluting the air masses that move across their areas and into the neighboring states.

To state the problem succinctly, state governments feel that they are better qualified to deal with local problems than the federal government, whereas the federal government feels that many problems are truly not local, so they require a regional perspective to deal with them adequately. Both positions are correct, so we are back to the question of political power, and both levels of government guard their prerogatives jealously.

Air pollution provides a good example of the difference between local and regional control. In the 1970s, states on the east coast were unable to meet the air quality standards of the Clean Air Act because pollutants emitted from factories and power plants in the Midwest were being carried eastward by the wind systems. Because New York could not force Indiana to reduce its emissions, the federal government had to become involved.

Indiana and other Midwestern states objected to the regulations imposed on it by the E.P.A. but had no choice but to comply. After all, policies adopted under "local control" were causing the problem. Indiana had been dealing with its own pollution problems by building very tall smokestacks so the pollutants would not settle locally, but instead, would be carried away by the winds. Out of sight, out of mind. Because states downwind of Indiana could not force changes in this kind of emissions policy, a regional approach was required, and so the provisions of the Clean Air Act were imposed on Indiana and other Midwestern states.

When local policies impose costs of various kinds on other jurisdictions, some sort of overarching authority is needed. That is why the first sentence of the U.S. Constitution contains the phrase "In order to form a more perfect union." The only way to deal with conflicts between members of a confederation is to have some entity with some well-defined (but not unlimited) authority over all of the members.

The right to develop and implement rules and regulations is not found explicitly in the Constitution, but it is implicit in the laws passed by Congress. The laws cannot be implemented, and compliance with them cannot be monitored, without the regulations developed by Executive agencies.

The reason we need regulations is the tendency for business enterprises to reduce expenses by among other things, discharging toxic materials into the environment. The drinking water in Flint, Michigan is a good example. Water quality regulations can only seem "burdensome" to someone who is indifferent to how his actions affect other people.

Economists use the word "externality" to represent harm done to someone else by your actions – that is, by an external source. Externalities involve the harm you cause to someone else, someone who has not given approval for you to victimize him, and who has not received any compensation for the "inconvenience."

Now think about emissions from an industrial source that fouls the air downwind from the facility. Everyone living downwind breathes materials that may be harmful and possibly even toxic. None of them have been consulted about the emissions, so they are being subjected to "externalities." The facility is discharging materials into a "common resource," the atmosphere, which may violate the Clean Air Act, depending on just what is being discharged. But just how it is violating the law depends on regulations developed by the E.P.A. which specify allowable levels of what can be in the emissions.

I don't know if this has ever been mentioned, but another way to justify regulations by Executive agencies is to invoke the "equal protection" clause of the Fourteenth Amendment. Briefly, those living downwind from the facility have the same rights as those living upwind from it.

The point that lobbyists for major industries do not mention is that free-markets do not protect the public from externalities. To focus only on the effect that regulations have on production costs is to ignore the social costs, which implies that the people affected are not as important as the profits made by the producers, even though the people do receive any of those profits, and they did not voluntarily relinquish any right to them.

When corporations claim that a regulation should not be imposed on them because of the cost of adhering to it, they are implying that society has no business meddling in their affairs. After all, they say that they are just trying to do what they are required to do, which is to make as much money as they can for their shareholders (In a sense, that is what bank robbers say too). So why is the government telling them they cannot do something because it imposes a cost of some sort on society? Which is more important?

Corporate management drinks the Kool Aid that convinces them that the nation's economy is independent of social mores. Their mission is to maximize the firm's assets, regardless of the collateral damage to society of some of their activities.

Their position is not always stated that bluntly, but it is implicit in their actions. But those who defend untrammeled corporate activities also seem to believe that corporations have a "right" to do what they do, and that the government should not be infringing on that right.

Those who argue that the economy suffers when government imposes restrictions on industry choose not to learn from history. Some benefits of regulations were discussed by the British political scientist Tony Judt, who wrote about social unrest in Great Britain in the 19th century because of unrestricted industrial practices. The conditions described in the novels of Dickens are certainly not anything we would want to put up with today. Facing social unrest, the government eventually instituted a number of policies that stabilized the lives of workers.

Those policies consisted of, among other things, upper limits of hours working in factories, a minimum age for child labor, regulations concerning working conditions, and legalization of labor unions. They all became a part of the social contract which largely eliminated social disruptions. Even though these policies increased the cost of operating factories, they were beneficial because they became sources of political stability.

Those regulations answered an important question, namely how was social upheaval to be prevented in a society that benefitted from exploiting a large class of low-paid and discontented people? Judt pointed out that the capitalist system in Great Britain benefitted from these regulatory mechanisms, regardless of their cost, because industrial operations work much more smoothly and profitably when there is little labor unrest.

The same question (about preventing social upheaval) was addressed in America between the World Wars. The rise of and legitimization of labor unions was instrumental in improving the lives of workers and in building the middle class in this country. But the concerted effort since the Reagan administration to destroy the union movement has moved the country in the opposite direction.

There is an important difference between conditions in America today and those in the U.K. in the mid-nineteenth century. Americans do not suffer from intolerable working conditions, partly because of regulations pertaining to workplace safety, and partly because many of their jobs have been moved overseas. A choice between bad working conditions or no jobs at all is not much of a choice.

Those responsible for today's conditions (the component of the population referred to as the "one-percent") are able to profit from the destruction of the labor union movement because the middle class in this country has forgotten why it exists.

My generation heard horror stories from our parents about what life was like during the Depression, but it seems that we did not pass those stories on, so subsequent generations are not aware of the value of a social contract.

The concept of "burdensome" regulations is the kind of propaganda we have been subjected to so often that many people believe it, even those who are breathing polluted air. But the logic behind the argument is flawed because it omits two things.

First, the specter of inflation does not really worry anyone who complains about the cost of regulations. What they are concerned about is the likelihood that they will not be able to raise prices enough to cover the cost of compliance, so the profits of their companies will drop – as will the share price on the stock exchange, which will make borrowing money to cover operating expenses more expensive. And shareholder dividends might have to be cut.

But the important flaw in the argument involves ignoring the fact that regulations are intended to help in some way the entire population. Polluted air causes health problems downwind from industrial facilities. The cost to society of this effect is not easy to calculate, but the fact that the population affected can be quite large means that the costs are not trivial.

Industries want the privilege of polluting the air other people breathe. They want the government to agree that they own the atmosphere we all need to survive. The cost to society for peoples' health means nothing to them.

The fact that the propaganda involving "burdensome" regulations is effective is remarkable because unlike the controversy over teaching evolution in the public schools, the importance of pollution is something that everyone in society can recognize. You do not have to be a chemist to know when the water coming out of your kitchen sink smells or tastes oddly. You know when some facility "upwind" from your home is releasing extraneous materials into the air you breathe. But unless the pollution affects me personally, am I likely to be concerned when I hear that a factory may have to close because complying with air quality regulations is too expensive? Or will I grumble that many local people will lose their jobs because of the regulations?

Environmental regulations are not examples of governmental "overreach;" they are attempts to protect the health of the entire public, based on scientific studies of pollution and its effects on people and on the environment in general.

I will say that another way. Environmental regulations are imposed to protect the public from pollution. If you accept that statement and still complain about the cost of the regulations, you are saying people are not as important as corporate profits. Few people are that blatant. More sophisticated anti-regulation people hem and haw and say that the extent of the health hazard from industrial emissions is not clearly known, and until it is, it is counter-productive to regulate the emissions. That says the same thing. To demand that unequivocal evidence of harm is needed before emissions should be regulated is to say that profits are more important than people. After all, why not say that emissions should be subjected to strong regulations until we know unequivocally the health effects they cause? Then we can adjust the regulations. I don't hear lobbyists for industries making that suggestion.

It is difficult to avoid the conclusion that those who use excuses based on economics have little to no regard for the health of the population. They worship at the altar of the free-market, which is equivalent to saying that they are practicing paganism.

Now scale what I've said up to consider the effects of emissions on the health of the population of the entire planet by thinking about the so-called "burdensome" regulations intended to restrict carbon emissions that contribute to global warming and climate change.

CLIMATE DENIAL

It is time for another shift in thinking, but not a change in what is being worshiped. The change is only in the scale of the phenomena being considered and the types of scientific studies that are being criticized. This time the resistance is not from those who object to *any* kind of regulations, but from one specific group, called "climate-change deniers," or just "climate-deniers," who argue that global warming and the resulting climate change are either not occurring, or that if they are occurring they present no important threats to society, or that until much more information is available about them, it would not be wise to make significant societal changes (such as curtailing the use of fossil fuels) because the changes would destroy the economy and plunge us all into poverty.

The climate-deniers criticize any regulations that would cause major reductions in the use of fossil fuels, which drive our industrial economy, regardless of the potential to cause serious societal disruptions due to climatic changes that will make some parts of the planet uninhabitable. They are afraid that curtailing the use of fossil fuels will cause the GDP to drop to a value near zero.

If climate scientists seem to be saying that the sky will fall if we do not reduce our use of carbon emitting sources of power, the climate-deniers are saying that the sky will fall if we do. Even if both groups are correct, the economic disaster predicted by climate deniers is not as certain as the disaster which will occur if we do nothing.

As before, we have two belief systems. One of them involves people who believe the results of scientific studies because they accept the idea that inquiry into the causes of phenomena represent the best way to learn about what is happening around us. The other group involves people who object to implementing recommendations based on the results of scientific studies about climate because they are aware that accepting the results will lead to curtailing the use of fossil fuels. The effect of that action would require major changes in the nation's and the world's economy, something they are not willing to consider.

At the very least, those changes would involve severe restrictions on the carbon emissions from power plants, which would force the electric power industry to utilize alternative sources of energy. That would bankrupt the coal and petroleum industries, which provide funding for a number of Conservative think tanks (which provide "experts" who claim that nothing needs to be done). By couching their resistance in terms of free-market economics, they are essentially practicing a quasi-religion. In fact, many of the beliefs associated with free-market economics have as little substance as some of the stories told in the Old Testament. And some of the people who reject the concept of climate change would rather base their beliefs on a literal acceptance of what consists of distortions of Economics 101 than accept the thinking behind scientific studies of the climate. So, as do the Creationists, this "quasi-religious" contingent rejects any scientific findings that contradict its dogma.

This attitude is an example of Confirmation Bias, in which the only kinds of information people accept as reliable are those that conform to their preconceived notions. Now, it is true that scientists are subject to Confirmation Bias too, but the fact that they base their studies on the basic laws of nature, instead of traditional beliefs that are not supported by evidence, reduces the likelihood of bias.

In our society, anything can be a commodity. Historic buildings, national parks, wilderness areas, drinking water sources, spawning grounds for endangered fish species, habitats for animals on the verge of extinction, and indeed, the very air we all breathe. All of these "things" are usually classified as "common resources," in the sense that regardless of local ideas about property rights, no individual can own them, so no one should have the right to exploit or damage them.

Common resources are a temptation for some people because no one is making money from the things, so they seem to be "wasted." Just sitting there, waiting to be used in some profit-making activity, regardless of whether that activity uses them up or permanently changes them. Cruise ships dumping raw sewage into the ocean, and industrial sites emitting particulate materials into the atmosphere, are examples of both exploitation and damage to common resources. In those examples, as long as the volumes are not too large for biochemical degradation processes to convert the emissions to harmless forms, little damage is done. But when the activities are such that long-term, sometimes permanent damage is possible (Lake Erie was considered "dead" at one time), society suffers because of the activities of a few people doing things they should not be doing.

That last thought pertains to the activities that are causing Global Warming and the resulting changes in climate, things that affect negatively everyone on the planet. The entire planet is the common resource that is being endangered by the use of fossil fuels to power society (or at least, societies in the developed world), and the need to curtail those activities is resisted by the climate-deniers because the restrictions involved would require significant changes in how society "runs its engines."

That explains why the climate-deniers lobby governments to ignore the ramifications of the activities of the energy industries; industries which exist to extract and use fossil fuels, even though the use of those fuels is endangering the common resource we call our planet. Those industries will be out of business once society realizes that their products are not competitive to the alternate energy sources that now are available, but which are not yet competitive on price.

So the belief system of the climate-deniers involves the desire to live in a society with as few restrictions from governments as possible, one in which resources can be exploited even if their use endangers the common resource we call our planet.

Creationists base their opposition to biological science on a literal interpretation of the Bible, whereas climate-deniers base their opposition to climate science on their insistence that our economy cannot function as it should if any restrictions are placed on market activities (e.g. restricting or curtailing the use of fossil fuels) in order to reduce the effects of global warming.

The strategy used by "climate-deniers" is similar to that used by Creationists, in that when they cannot prove that something is wrong (which is nearly all of the time), they resort to questioning its reliability. They *have* to resort to that strategy because if they were to concede that the scientific findings are believable, and accurate, and ominous, they would have to accept the need for regulations that restrict, and eventually prohibit the burning of fossil fuels. Because restrictions are anathema to them and because funding for the organizations associated with climate-denial comes largely from the fossil fuel industry, they cannot accept the scientific findings.

There should not be a controversy over global warming and climate change, but one does exist, one which has been manufactured by the fossil fuel industries and the "think tanks" they fund. The principles underlying the subject are straightforward enough that they can be taught in middle-school science classes. And I hope that they are. But our society may not have enough time to prevent catastrophe if we wait for those teenagers to become adults in decision-making positions. So it is important to convince today's adults that the matter is sufficiently important that significant changes need to be made as soon as possible, regardless of how those changes will affect the "bottom lines" of the companies that extract and sell the fossil fuels.

The science underlying climate change is sound. Here it is, without using the crayons that Republican politicians seem to need.

The atmosphere contains small amounts of several gases that retain heat, so they slow the escape of heat radiation to space. Every day we put more of one of those gases into the atmosphere. It follows that more heat will be retained in the atmosphere, which will cause its temperature to rise. How clear is that?

It also follows that in addition to the entire planet becoming warmer, the extra heat will stir up the fluid we call the atmosphere, resulting in more extreme weather-related consequences, but explaining why those things will happen may require some crayons.

Based on the explanation just presented, it is clear that to understand global warming and climate change, we need to know something about the temperature of the atmosphere, the gases in the atmosphere, and the reason why some gases affect the global temperature, namely the carbon emissions from industrial sources that put more of one of those gases in the atmosphere.

Reliable thermometers were developed in the 19th century, and extensive sets of measurements of temperature have been made since 1880, both on land and at sea. When the temperature data set is compiled and a graph showing the average world-wide temperature over time is displayed, we see that the average temperature of the planet has been increasing markedly since 1880.

The values are average ones, so complaints of the form "It's cold today, so the planet cannot be warming," are specious. The "average" value of a set of numbers is based on some that are larger and others that are smaller.

Now, is the increase in temperature since 1880 important? Does it really matter? The answer is yes, it does matter, because a number of proxy methods to measure temperature in the distant past exist, which indicate that the increase we observe is unusual because of how rapidly it is occurring. In fact, if we consider the rate at which temperature is increasing as well as the actual values, the period since 1880 is quite unusual. So it makes sense to figure out why today's conditions differ so much from past conditions.

That is how scientists work. They see something and say: "That's funny. I wonder why that is happening." To answer that question, we start with the *Greenhouse Effect*.

The words *Greenhouse Effect* refer to the ability of several gases in the atmosphere to temporarily trap heat that should be radiating out to space. Energy from the Sun strikes the planet and warms the land and oceans, which in turn send the energy back out to space. If what goes out matches what comes in, the temperature of the planet remains constant. But the atmosphere affects the balance of input versus output. Most of what comes in (what we call sunlight) vibrates vigorously, at a fairly high frequency. But what goes out vibrates at a lower frequency.

A law called the Stefan-Boltzmann law relates the frequency of radiation to the temperature of the source. The higher the temperature, the higher the frequency. The temperature of the surface of the Sun is about 5,000° C, whereas that of the Earth is about 12° C (about 55° F). So, the law says that the radiation from the Sun vibrates very vigorously. Therefore, most of what strikes the Earth's surface vibrates vigorously, whereas that emitted by the Earth vibrates much more slowly, in the range we call infra-red, or "heat" radiation.

Several gases in the atmosphere absorb some of that heat temporarily, preventing all of it from leaving immediately. So the atmosphere is warmer than it would be if those gases were not present. About 60° F warmer. As mentioned above, the average temperature of the atmosphere is about 55° F, so without the gases mentioned, that temperature would be about minus 5° F, in which case there would be no liquid water (and therefore, no life as we know it) on the planet. This heat-trapping process is called the Greenhouse Effect, because it seems to be analogous to what goes on in a greenhouse. So, the greenhouse gases keep the planet livable by maintaining a range of temperatures (whose average is about 55° F) that are compatible with the presence of life.

Now, if things usually are balanced, why has the average temperature of the planet been rising for the last 200 or so years? Around the turn of the 20th century, a Swedish physicist named Arrhenius pointed out that for most of the previous century, industrial activities relied on burning coal for power, which released Carbon Dioxide (CO_2) into the atmosphere, and that gas was known to absorb heat. He predicted that the planet's temperature would increase if those activities persisted.

Well, they did persist, and from time to time I see a graph showing how the concentration of the gas has varied with time, plotted alongside that of the average temperature variations. Both of the lines are rising with time, so the graphs suggest a relationship between the two sets of data. The question is does one (the gas) affect the other (the temperature), as Arrhenius predicted? The answer to the question is what is behind the controversy over global warming and climate change.

Now let's pause for a bit and examine how the climate-deniers respond to the kind of graph mentioned in the last paragraph. They cannot deny that the temperature has been rising, so to rationalize what the data seem to indicate, they have to explain why the increase is not important. They point out that temperatures have always risen and fallen, so the warming of the planet over the last two centuries could be due to a "natural" fluctuation of the climate, and could have nothing to do with the amount of Carbon Dioxide in the atmosphere.

There are three ways to counter the suggestion that a natural fluctuation could be causing the observed warming. First, no natural system, such as the climate, changes spontaneously. There has to be a cause for any change. And only two possible "natural" causes are possible. The warming observed since the 19th century could be due to an increase in the energy released by the Sun, but the Sun is under observation, and no increase in its output has been measured. A second possibility is that volcanic activity could be the cause, but there has been no major increase in eruptions lately either. Everything that happens must have a cause, so unless someone thinks of a plausible natural cause for the warming, the suggestion that it is due to a natural fluctuation need not be considered.

A third reason to reject the "natural fluctuation" suggestion pertains to the gas Carbon Dioxide (CO_2). This gas is one of the "greenhouse gases," which means that it absorbs some of the heat radiated out into space by the Earth. Without these gases, the planet's average temperature would be about 60° lower it is, so no liquid water would exist here, and life as we know it would never have developed. The greenhouse gases make the planet habitable.

Since the 19th century, the concentration of CO_2 in the atmosphere has increased by about 40% due to the burning of fossil fuels (there is no other explanation for that increase), and half of that increase has occurred since 1970.

Now consider this question. If the concentration of a gas that traps heat has increased appreciably over the last two centuries (while warming was occurring), what has it been doing, if it has not been contributing to the warming? The climate deniers cannot answer that question.

When there is a way to explain something, and there are no realistic alternatives, a sensible person accepts the explanation. I know of no explanation from a climate denier that deals with this question. "What has the gas been doing?"

Let's add some details to the discussion. Here is a fact. The Greenhouse Effect is caused by the presence of five gases in the atmosphere, which temporarily absorb the heat emitted by the Earth before it can get back out to space. They are Carbon Dioxide (CO_2), Methane, Water Vapor, Nitrous Oxide and Ozone. That fact has been verified by numerous laboratory measurements. There are some other heat-absorbing gases in the atmosphere, but they are present in very small amounts, so they are not as important as the five mentioned above.

The presence of these gases results in a warm layer of air near the planet's surface. They act like a blanket because they absorb the heat trying to go back out to space. Without this effect, the average temperature on the planet would be lower, because that heat would go out to space. In fact, the average temperature would be about 60° F colder than it is now.

About minus 5° F instead of plus 55° F. In case you have forgotten, water freezes at plus 32° F, so without those gases in the atmosphere, no liquid water could exist on the planet, which means there would be no life as we know it.

The effect I've just described is called the Greenhouse Effect because it is said to be analogous to the reason a greenhouse tends to be warmer than the area outside the structure in the winter. That is not quite what happens, but it is close enough that I won't take up space explaining the difference.

If there are five Greenhouse Gases in the atmosphere, why do we hear only about Carbon Dioxide (CO_2)? Because it is the only one over which we have any control, and it is the only one that has been increasing markedly in the last two hundred years. And it is increasing because we are burning large amounts of coal and oil to power our economy. The amount of CO_2 in the atmosphere has increased by about 40% in the last 200 years, and half of that increase has occurred since 1970.

Now let's repeat the question, because it cannot be repeated too often. If a 40% increase in a gas that absorbs heat has not contributed to the higher temperatures that have been measured, what has that gas been doing?

Now let's put the discussion in context by mentioning another analogy from basic Physics.

The momentum of an object is defined as its mass times its velocity. A bowling ball has more momentum than a volleyball rolling at the same speed, so the physical properties of a system are important, as is the rate at which it is moving. That means that if you want to stop something that is undergoing change, you have to address both components – its size and weight, as well as the rate at which it is changing. It is harder to stop the bowling ball than the volleyball, unless the bowling ball is barely moving and the volleyball is moving very quickly.

These comments are not limited to rolling balls and vibrating objects, which are the kinds of things studied in an introductory Physics class. But these principles are the reason climate scientists keep warning us that something must be done about carbon emissions, and that it must be done soon. The rate at which carbon dioxide is being released to the atmosphere is increasing (thereby warming the atmosphere and triggering extreme weather events), even though some of the gas is being absorbed by the oceans (thereby making the waters more acidic, which affects marine life). And the cause of these changes is known.

This is not a situation where we have the luxury of waiting to see how bad something gets, and then deciding to address the problem. The consequence of Newton's work in the 17th century was that the more rapidly systems change the harder it is to stop the changes. But that lesson is being ignored today.

Arundhati Roy, writing about the devastation in the lives of people in India who are displaced by the construction of large-dam projects, mentioned an adage used by people fighting construction of the projects. "You can wake someone who's sleeping. But you can't wake someone who's pretending to sleep." People who discount the consequences of the projects in India use a variety of excuses – national development, progress, etc. But they don't want to talk about what is happening to those who are displaced by the projects. They pretend they do not understand the real issues.

Now apply the adage to climate change. Those who deny that global warming is happening, or if it is, its effects will be beneficial, are not sleeping; they are just pretending to be sleeping. The professional "deniers," who sometimes testify at Congressional hearings and at conferences at right-wing think tanks, are funded by the energy industries. They usually are economists who warn that the economy will be devastated if the fossil fuels are "left in the ground." What they really mean is that the companies paying their salaries will be bankrupt. But they never get around to mentioning the migration of millions of people fleeing the consequences of climate change. For some reason, there is no way to fit that consequence into their arguments.

There is a difference between a reason and an excuse. Excuses are all we hear, because considering the importance of the topic, there is no rational explanation of why we should not leave all the fossil fuels in the ground. The "bottom line" of a coal company's annual report is an excuse, not a reason.

When discussing a controversial topic such as global warming, rational people base their discussions on evidence. The best evidence to support the claims of those who say that society must curtail its use of fossil fuels consists of data collected from a continuous core of ice extracted from an ice sheet in Antarctica. Studies of the gases in the pores of the ice provide information about the climate over the last 800,000 years. Think of that. A data set representing a time span of nearly a million years exists.

The amount of Carbon Dioxide (CO_2) in the pores of each thin slice from the core has been measured, as has the ratio of two isotopes of Oxygen (one "weighs" 16 units and the other 18 units), which provides a temperature gauge because the amount of the most common isotope (16) is constant, but the amount of the other changes with the temperature. Comparison of the two sets of measurements show that as the concentration of CO_2 has increased, the temperature of the atmosphere increased.

An important characteristic of the data set is that over the 800,000-year time span, the concentration of CO_2 has never exceeded 290 ppm (parts per million). But today, the concentration is at about 400 ppm, which is 38% higher than at any other time recorded in the ice core. And those high values have all occurred since the beginning of the Industrial Revolution, early in the 19th century. Note: From time to time I may refer to "about a 40% increase," rather than 38%, because round numbers are easier to remember.

So, what do we have here? First, there is a correlation between CO2 and temperature. Second, although the CO_2 values (and therefore the temperature) have varied over the last 800,000 years, they have never been as high as they are now. Third, the current value (nearly 400 ppm) developed abruptly - in the last two centuries.

Never before has anything produced CO_2 that quickly. Nothing. So what is so different now, that the values shot up so quickly? The only explanation is the Industrial Revolution, which has been possible because society has been burning coal to produce power.

Keep in mind that coal is nearly pure carbon. Even "dirty" coal which has sulfur in it is mostly carbon. When you burn something, you let it react with oxygen (that is what "burning" means), so burning coal involves combining a carbon atom with two oxygen atoms (the carbon atom has four "open" sites to which the two in each oxygen atom can link), so you get CO_2. There is no way to burn coal without producing CO_2. That is not something that technology can prevent: it involves laws of chemistry.

Remember what Marcia McNutt said: Science provides a way to decide whether we should believe something.

Now, there are not enough plants on the planet to absorb the CO_2 we emit every year, so most of it sticks around in the atmosphere and traps heat, some of which warms the atmosphere and some of which warms the oceans. And the concentration of CO_2 is continuing to increase, so it is reasonable to assume that the Earth's average temperature is going to continue to increase, as will all of the negative effects that warming produces, that are grouped together under the rubric, Climate Change.

That is, the evidence shows that the planet has begun warming in the last 200 years much more rapidly than has ever occurred previously. The relative time span (800,000 years divided by 200 years) is equivalent to a line with a slope of 4,000. A slope of "one" is equivalent to a 45° angle, so imagine jogging up a slight grade and then running into a nearly vertical wall.

So when climate deniers say that fluctuations in climate have always occurred, they are correct. But what is going on now completely dwarfs the minor fluctuations that have occurred in the last 800,000 years, so their argument is worthless. The climate deniers have no evidence to support their position.

The data from the last 200 years provide a "smoking gun."

The rules restricting carbon emissions from power plants that the E.P.A. finalized in 2016 were not a surprise; they are really a reflection of what has been known to be necessary for about a decade. But as expected, the governors and senators of several states reacted negatively to the announcement that the rules were going into effect. Because Congress would not take any action on the matter, the E.P.A. used the authority to develop regulations granted to it in the Clean Air Act, which was approved by the Supreme Court. The Court agreed with Appeals Court rulings, which said that Carbon Dioxide must be (not *can be, must be*) regulated under the Clean Air Act. Note that this is the "Roberts Court," which is not known to be a hot-bed of liberalism.

The limits vary from state to state, with the average being a 30% reduction by the year 2030. To listen to the governors, one would think that the sky is falling, but they somehow managed to avoid mentioning that 10 states had already reduced carbon emissions by that amount or more, as of 2012, and several other states are well on the way to doing so - well before the deadline.

The limits assigned to each state by the new rules vary because some states have few options; they have few or no power plants that can burn natural gas, and thereby reduce their dependence on coal. According to a map I have seen, Indiana has made no progress in reducing carbon emissions, and because of the economic importance of the coal industry in the southern part of the state, our legislators have decided that they "will not go gently into that dark night."

But notwithstanding the politicians, from whom we should expect histrionics, a number of utility executives see no problem in satisfying the rules. In the Northeast, a cap-and-trade system has been implemented, so utilities that find ways to become more efficient sell their emission permits to less efficient companies. The important part of the system is that from time to time the cap (the total amount allowed in the region) is lowered.

Here is one case in which the free market does indeed reward innovation and forward-thinking. The less carbon a company emits, the more money it can make by selling its emission permits.

To add salt to the wounds of the Mid-West governors, the states that have already implemented the reductions have seen their economies rise by more than the states which have not implemented reductions, so the rules are not likely to have the destructive impact the Republicans would have us believe. It seems that the dire predictions from politicians and professional climate deniers are little more than what they feel is required to maintain their conservative credentials. If you don't pander to your base you might not be re-elected.

Of course, the recommended limits are not nearly enough to have a noticeable effect on climate change, because they are coming at least a decade too late. But their effect is intended to be political, in the sense that the rules give the administration leverage when dealing with other countries that are reluctant to start similar reductions.

Those who choose to deny the existence of things they do not like are really saying something about themselves, rather than about the nature of reality. The "climate change deniers" have a hard row to hoe because the evidence for the change is not only strong, it makes sense logically.

One of the most common statements people make about scientific findings they do not like is to say that it "is only a theory." That response "has legs," because most people do not know that the word theory does not always represent ideas that are problematical. In fact, scientists use the word to represent an explanation of a phenomenon that is unequivocally sound.

For example, consider Newton's "theory of gravity." Among other things, it explains why things fall down instead of up when you release them; it explains why the planets orbit the Sun in elliptical orbits; it explains why the Earth is not a perfect sphere, and it explains the origin of the ocean tides. No other hypothesis explains so many different phenomena simultaneously, so Newton's equation for the force of gravity, along with his equations of motion are said to constitute his "theory of gravity."

There is nothing equivocal about it. NASA sent men to the Moon based on it, and the orbit of every satellite that sends gps signals to your smart phone and television signals to ground-based receivers is based on it. As one more bit of evidence for the validity of this "theory," consider that no "denier" is willing to step off the roof of a tall building to prove that there is some uncertainty in its predictions because "it is only a theory."

So, to a scientist, a "theory" provides the best explanation for a phenomenon because it explains more aspects of it than any other possibility.

As was already mentioned, one of the ways "climate-change deniers" work is to try to convince the public that even if global warming is occurring (and it *is* occurring), it could be a "natural" change, and not due to human activities associated with burning fossil fuels (mostly coal). They do not suggest what could be causing this "natural" change, because no "natural" cause is known that could produce the large increase in the planet's temperature that has occurred in as short a time as the last two hundred years.

What none of the "climate deniers" mention is that in addition to the very rapid rise in carbon dioxide in the last 200 years, which does not happen "naturally," there is another way to demonstrate that the increase in carbon dioxide in the atmosphere is due mainly to burning coal to generate electricity. It involves two different stable forms of carbon: Carbon 12 and Carbon 13. Most of the carbon on the planet consists of C(12), but about one percent of it consists of C(13), which weighs a bit more because it has an extra neutron in its nucleus. Note: C(13) is not the unstable form of carbon which is used in radioactive dating studies. That is Carbon 14. C(13) is stable.

When animals exhale (the process is called "respiration"), they (we) exhale a mixture of the two forms of carbon – about 99% of it is C(12) and the rest is C(13). But when plants absorb carbon dioxide from the atmosphere, they prefer C(12), so the effect of vegetative growth is to increase the percent of C(13) in the air relative to that of C(12). That is, as plants remove C(12) to grow, they leave the C(13) behind, so the ratio 13/12 increases. Keep in mind that coal has essentially no C(13) in it because it is formed by the decomposition of plant material (and plants prefer the lighter form). So when we burn coal, we add mostly C(12), and the ratio 13/12 decreases.

The fact that burning coal puts C(12) in the atmosphere rather than C(13) provides a "fingerprint" that is pertinent to global warming. The more coal we burn, the lower the amount of 13 relative to 12, *and that decrease is observed over time* in the carbon dioxide in the atmosphere.

Let me repeat that. The ratio of 13 to12 in the atmosphere is decreasing, which indicates that a substantial amount of the carbon dioxide in the atmosphere is C(12) from fossil fuels. That is an indication (proof?) that the increase in carbon dioxide and therefore in global warming, is due to human activities - namely the use of fossil fuels to power our society. There is no other way to explain the observed change in the ratio of the forms of carbon.

This data set is another "smoking gun" (in addition to the very sharp increase in CO2 and temperature). There is no point in looking for natural causes for the increase in the planet's temperature. It will be fruitless.

To understand why it important to worry about what we are doing to the environment, consider the fact that for the first billion years of the planet's history, the only forms of life were structurally simple single-celled organisms (bacteria). They were "anaerobic" organisms (they did not need oxygen) but that was fine because there was no oxygen in the atmosphere.

Over the next two billion years, slightly more complex single-celled organisms developed and dominated.

Pause here for a second to think about this. For three billion years, the planet got along just fine without any complicated life forms. Is that good for your ego? Do you still feel that everything we see around us was intended for our use?

Okay, back to the point. The more complex organisms which developed were able to use sunlight for energy and they released oxygen as a waste product, so over that two billion years, oxygen accumulated in the atmosphere.

By about a billion years ago, the amount of oxygen in the atmosphere was enough to allow multi-celled organisms which used oxygen for energy to develop. And by one-half billion years ago, organisms complex enough for us to call them "animals" had developed.

Fast forward to about 100,000 years ago, which is when modern humans appeared. A species that has existed for only the last one ten thousandth of the billion years that complex life forms have existed thinks it is special. Forget it. We are the new kids on the block, folks. Nearly everything that lived prior to our appearance is gone - for a variety of reasons, but usually because of widespread environmental changes.

Now consider this. If numerous species became extinct mainly because their habitats were destroyed, for one reason or another, how much longer can our species survive if we keep destroying the environment in which we and everything else live?

Gaia is a tough bitch. She doesn't need us. A system that functioned for billions of years without us will get along just fine when we are gone, even if we screw things up so royally that it has to start over with single-celled organisms. Until we understand that, there is not much hope for the long-term survival of our species.

How can we evaluate the contrasting claims made about the research done by climate scientists? The right-wing media pundits seem to think that these professionals don't know what they are doing. Is that really likely? How would professional talk-show hosts know? Or, as some of the climate "deniers" claim, is it likely that the predictions made by climate scientists are self-serving, in that they are designed to generate more research funds for more studies? That is, some of them claim is that the climate scientists are shouting "the sky is falling" in order to maintain their jobs. That might be possible if a small number of the scientists were making the predictions. But it is hard to imagine a world-wide conspiracy intended to generate more research funds when thousands of scientists, working in different kinds of organizations, in different countries, being funded in different ways, make the claims. In addition, many of the climate scientists have tenure at universities. If you are not familiar with that term, "tenure" is a "no-cut" contract.

But that kind of security is not enjoyed by most of the climate deniers, who are either economists working at Conservative think tanks that are funded by the energy industries, or are employed directly by companies in the energy industries. Their security depends on funding from the energy industry. There is no reason to believe that any of these economists have any training in atmospheric science, and clearly, those whose jobs and incomes would be threatened by reducing our dependence on fossil fuels have an incentive to discount the work of the thousands of climate scientists who have been collecting evidence of global warming and its relationship to carbon emissions for decades.

So, any time you hear media pundits criticizing climate scientists for not being impartial or neutral, keep in mind that the pundits' sources get their salaries directly from the companies that are pumping millions of tons of carbon dioxide into the atmosphere every year. Can anyone believe that those sources are impartial?

And can anyone believe that the media pundits are not pimping for the energy industries that quite likely advertise on their programs?

I should add one thing here. There are, in fact, a few climate deniers who are atmospheric scientists. About a handful, as opposed to the thousands who contend that global warming is causing climate change. A handful versus several thousand. Which group should we believe? I am reminded of a quip published sometime in the 1930s by the sportswriter Grantland Rice, who adapted a thought in Ecclesiastes, and said:

The race does not always go to the swift,
Nor the fight to the strong.
But that's the way to bet.

In other words, when you bet on a long shot, you usually lose. And this wager is one that we need to win.

In the introduction to one of his books on the Gaia Hypothesis, James Lovelock said that the Gaia Theory may be wholly or partially in error, but to a real scientist that is not as important as how well the theory fits certain criteria. His criteria were: Is it useful? Does it suggest interesting experiments? Does it explain the data that have been gathered? What are its predictions? Does it have a mathematical basis?"

These are the kinds of criteria considered in a valid scientific investigation. Objections based on economics or practical politics may be valid in their own domains, but they have nothing to do with the science involved. When a scientific hypothesis satisfies the relevant criteria, it matters not if its conclusions are practical or if recommendations based on them are politically possible to implement.

And it does not matter if at some time in the future, further studies may require that the ideas be modified because new information becomes available. What matters is that if what is known today satisfies criteria that the preponderance of professionals in the discipline accept, the conclusions of the studies are worthy of acceptance by intelligent people. The conclusions "will do" until someone shows that they need adjustment. That is all that we can do.

In some cases, someone will object that provisional acceptance is not enough; that certainty is required when the conclusions and recommendations call for large-scale social and political changes, such as would be involved say, in curtailing the use of fossil fuels to prevent widespread devastation due to climate change. In such cases, all we can say in response to them is that many political decisions are made that are based on the flimsiest evidence (declarations of war, for example), so why should decisions that will impact the entire population of the world require absolute certainty, even if that were possible?

Politicians usually think their goals are important and that the purpose of a scientific study is merely to validate those goals. But they are not acting sensibly when they think that. If the results of a rationally conducted study suggest that a policy *must* be implemented to avoid dire consequences, any politician who objects should have a better reason than to say that he wants to be more cautious. Otherwise, he is saying more about himself than about the study or those who conducted it.

Based on findings published in the media on a regular basis, how can a sensible person not decide that things such as global warming, species extinctions, and ocean acidification are issues that need not be dealt with? If you are sensible, you will realize that they really are warnings that society has reached the point that we can make permanent changes to the environment in which we live. And some commentators feel that we are on the verge of a tipping point, from which there may be no return.

Much of the resistance to implementing policies that would reduce if not prevent the effects of what these warnings are telling us is due to some unwarranted assumptions about the ability to deal with problems as they arise. Some people have too much confidence in engineering solutions. Some of them probably also believe in the tooth fairy.

And the number of future "issues" is unpredictable because the atmosphere-ocean-life system is highly non-linear so we cannot predict ahead of time what kinds of things will occur when we exceed some threshold. In fact, we do not even know what kinds of thresholds are likely to emerge, much less what their "critical" values are likely to be.

In addition, the farther we go down the road of tampering with natural systems, the less likely we will be able to stop some practices in time to prevent crossing such a threshold. The human "footprint" is sufficiently large that it is resistant to being downsized, and self-restraint is not a characteristic of our species. Consider the reluctance to consider leaving fossil fuels in the ground because that would have a very large negative effect on the economy.

Anyone with a grain of common sense should know that continuing to burn fossil fuels will increase the planets' average temperature, which will exacerbate climatic changes, none of which will be beneficial. The planet will survive but after the effects of the changes are felt, society will not likely resemble what we have today.

One of the reasons there is resistance to trying to deal with the problems that will develop involves time scales. The major effects of climate change could take a generation or two to develop, whereas implementing policies to prevent those changes will have an immediate effect on the world's economy.

We are not used to taking the "long view," so we may end up watching disasters occur because we were not willing to take steps to prevent them when it was possible.

There seem to be lots of people who either are oblivious to the consequences of global warming and climate change, or are convinced that somehow, solutions will be developed that will prevent the direst consequences from happening. After all, some people say, we managed to avoid nuclear annihilation during the Cold War. All it took was the resolve not to push "the button." With that resolve, it is as if the problem never existed.

Unfortunately, when we think seriously about climate change, we realize that there are many buttons, and deciding to avoid pushing one of them does nothing about the others. The consequences of global warming are so many and so varied, that there is no single organization or nation with the ability or authority to deal with all of them.

In one of her talks, Naomi Klein pointed out that global warming is telling us that our civilization is so complex and technologically developed that we can produce changes in natural systems on a global scale. If we continue to ignore this message, our society will constantly be faced with looming disasters. The very presence of those disasters will indicate that we are determined to continue to deal with effects rather than causes. That is a mug's game.

The attitude that we must deal with problems that arise because of our activities instead of doing what is necessary to prevent them from happening in the first place is a sign that we are ignoring what Ecologists have been trying to tell us. We think we are the end-product of Evolution (at least, those who will acknowledge the existence of Evolution); that we are the species that was "meant to be," rather than just another component of a complex ecosystem, whose existence (that of the ecosystem) is the result of a few billion years of "experimentation" driven by geologic processes and natural selection. That kind of attitude gives people a false sense of security.

The only sensible way to deal with problems is to avoid doing the things that cause them. But that will require a major change in the way we think about ourselves and about the world in which we live. To survive, we cannot continue building metaphorical walls or moats that give the appearance that we are separate from nature. How that change in thinking is to be done is not immediately apparent, but it is what needs to be done.

The group of people called "climate deniers" refuse to believe that the average temperature on the planet is increasing and that the cause is the amount of fossil fuels we burn every year, which puts more and more carbon dioxide in the atmosphere.

The "deniers" say that the computer models used by climate scientists are too simple, and therefore cannot be trusted to be accurate, because they do not represent everything that affects climate. So, they say policies aimed at curtailing carbon emissions, which will devastate the economy, should not be based on modeling studies. Of course, those doing the modeling studies know that the models are not completely realistic, but, unlike the deniers, they also know the extent to which the results of their studies are reliable. So let's review what the models are and what they do.

Climate models are computer programs that solve a set of equations which govern the heat balance on the planet. One equation represents solar energy striking the planet, another represents the heat energy emitted back to space by the planet, another represents the movement of air in the atmosphere, and another represents the transfer of energy between the atmosphere and the oceans. The equations were not created to represent the climate; they were developed in the 19th and early 20th centuries in studies of different kinds of physical problems involving radiation and fluid mechanics. That was long before climate was an issue studied by scientists. Climate studies are just one application of the equations.

Since WWII, digital computers have been used to address physics problems whose governing equations could not be solved the traditional ways, with "pencil and paper." The equations are linked together in a computer program in which the algebraic expressions are converted into a form suitable for arithmetic calculations. The computers do arithmetic very quickly, so it is feasible to use them to address problems that are seemingly intractable.

Several models exist, developed at different climate science centers, but they all have some things in common. They use the same basic equations, differing only in how they represent the small-scale features, and they get largely the same results, in that the output of any of these computer models represents the basic features of the climate. When started up with data available from the early 19th century and run to the present, the models' predictions for today's climate are similar and largely accurate. By that I mean that they give average temperatures that match observations fairly well. In addition, they predict other things, such as the existence of El Nino events, although the timing of the events can be off. That timing is not really a problem, because it is remarkable that they come up with these phenomena at all. The fact is that they represent the gross features of today's conditions accurately.

The models provide a "picture" of climate that is representative of *what* we experience on average, albeit not always *when* we experience it.

They work much better than models of the economy because they are based on physical principles, whereas the economic models are based on previous observations of economic booms and busts.

Of course, there are lots of things the models cannot represent. Sometimes that is because we do not completely understand what is actually going on in the field. For example, shifting of ice sheets in the Arctic and Antarctic are an important feature which affects sea levels that have not yet been incorporated into the models.

But the fact that they can represent what is going on today should convince anyone with an open mind that they should be able to represent average conditions that will develop in the next decade or two, in response to different policy decisions, such as a 20% reduction in carbon emissions. And this technique should also be able to indicate what degree of curtailment of carbon emissions will be needed to prevent the average temperature from exceeding a specified value in the next decade. That is why the models were developed. They represent a way to learn which options are likely to produce which results.

Anyone who objects to having policies developed on the basis of modeling studies is either completely scientifically illiterate or has other reasons for rejecting recommendations based on science. The fact that many politicians receive campaign contributions from companies in the energy industry comes to mind.

One of consequences of global warming is a shift in growing zones. As the planet warms, the zones in which plants can grow shift to the North. And the behavior of many animals also changes. Elizabeth Kolbert wrote a book on Climate Change, and in one of its chapters she interviewed a number of Biologists who have been studying changes in animal and plant behavior over the last few decades. She mentioned frogs in central New York which mate ten days earlier than in the past, and she noted that the date of peak blooming for spring-flowering plants in a Boston Arboretum advanced about eight days. In Costa Rica, some birds which used to concentrate in the lowlands have started to nest on mountain slopes, and in the Sierra Nevada of California, some butterflies are now found at elevations three hundred feet higher than one hundred years ago.

Kolbert pointed out that any of these changes could be explained by reference to purely local conditions, but the only explanation that makes sense of all of these changes (and many others) is global warming. Once again, a cause that explains many different phenomena simultaneously provides the best explanation, and is entitled to be called a "theory," in the scientific sense.

One of the nice thing about Kolbert's book (*Field Notes from a Catastrophe*) is that she considered little things, such as butterflies (and also mosquitoes) as well as big things, such as glaciers in Greenland.

Here are some points made by an Evangelical climate scientist who tries to overcome the apathy about global warming in the Evangelical community. The article I saw did not mention the kind of reaction she gets.

1) Conservation is conservative. What is more conservative than conserving our natural resources? Although most political Conservatives deplore big government, conservation does not require federal mandates. She noted that Texas produced about 40% of its energy needs from wind energy. Going further, she said that a map showing the greatest potential for wind energy production highlights the "red" states. So, a shift to this form of renewable energy would be consistent with the core beliefs of the residents, because it would represent a market-friendly approach to the production of energy. And wind-energy companies get far fewer government subsidies than do fossil-fuel companies.

2) To those who say God would not let us destroy ourselves, she responds, that in fact, God would let it happen. If you believe in free will, you should also believe that we have to live with the consequences of our choices.

3) Even though "end time" believers say that climate change does not matter because the second coming is imminent, it does matter, because that attitude is also equivalent to saying, "screw other people," which is not consistent with the New Testament. She

quotes Paul who said that we don't know when the second coming will occur, so we have to live our lives responsibly, and that means avoiding things that hurt other people.

4) When speaking to "Young Earth" believers about global warming, she shows temperature data from ice cores that go back no more than 6,000 years, so that even they cannot deny that the planet is warming.

5) Concerns about climate change are not just about the fate of frogs and other endangered species. Climate change affects people - adversely. For example, the threat to food supplies threatens the lives of millions of people. Christians cannot honestly ignore that.

6) If you believe that God created the world, why would you treat it like garbage?

I hope that the last point is how she ends her talks. It should stimulate thinking.

"Climate deniers" refuse to believe that global warming is caused by carbon emissions from burning fossil fuels because they do not want to change the nice cushy lives they lead. So they reject anything in modern science that confirms the link between fossil fuels and climate change.

Here are six sets of observations that are consistent with the concept of global warming. That is, these should be happening if global warming is occurring.

Cooling of the upper atmosphere, less heat escaping to space, more heat returning to Earth, winter warming faster than summer, nights warming faster than days, and a pattern of ocean warming.

In fact, these six things are all occurring, but this set of observations is more than just consistent with global warming. The set of observations cannot be explained by any other single process.

That last sentence is important.

There may be another explanation for cooling in the upper atmosphere, but that explanation is not likely to pertain to the ocean temperatures or the fact that nights are warming faster than days. The point is that Global Warming provides an explanation for all of them. The simplest, most comprehensive explanation. What more do we need?

So here we have an example of how to solve a problem. Find a possible solution, one that explains as much as possible, and then determine whether any other explanation exists that can explain the data set. If there is no other way to explain what you are studying, you can have confidence that your solution is good enough to be used, until an alternative is found that explains the same data and perhaps, more.

The Climate Science community routinely publishes the results of research on the effects of carbon emissions on the planet's average temperature and on dozens of aspects of different ecosystems that are threatened by global warming. Like all scientific publications, these reports contain extensive information on how the studies were done, the results of the studies and their consequences. I have seen lots of references to these reports in the media, many of them critical of the conclusions, but none of the critics indicate *why* any of the reports might be wrong (in the sense of inappropriate assumptions, biased data collection, flawed calculations, or logical errors).

In other words, the evidence collected by the professional scientists supports their conclusions that the planet is warming, and that the warming is having a variety of effects of global ecosystems, few if any of which are beneficial. They collected and analyzed data and then reported the results of their analyses. That is how scientists have been working for the last two hundred years. And no one has shown that or how the conclusions might be incorrect.

But the climate-denial community, funded by the energy industry, has managed to convince large numbers of people (especially politicians) that all of that work is flawed. The deniers have not tried to find errors in the reports; that is not necessary, because the general public does not read them and would not understand the papers if they did (especially the politicians). All the deniers have to do is suggest reasons that the entire effort is suspect. And they do that in a number of ways.

Nothing the deniers do involves any evidence. They cannot show that the climate scientists are wrong, so they obfuscate by suggesting that they are not perfect; that they have not included everything in their studies that could possibly affect climate; and that it would be prudent to wait until some "proof" is available that they are correct. Of course, it is not clear what kind of "proof" they would accept. Sometimes it seems that their masters in the energy industry would not be bothered if conditions here mirrored those on Venus, where the average temperature is about 800° F. because its atmosphere is mostly carbon dioxide.

The climate science studies are data-driven, whereas the climate deniers' response involves confusion and obfuscation. Which should people (especially politicians) believe? It seems reasonable to say that the minds of those (especially politicians) who don't believe the results of data-driven studies are completely closed.

If climate scientists seem to be modern-day Cassandras, always predicting gloom and doom, remember that it was foolish to ignore Cassandra, because she was always right.

CONCLUDING COMMENTS

There are many kinds of scientists, but they all share one characteristic: they believe things that are supported by evidence. And that evidence must be objective, in the sense that if several people collect it independently, they all get the same results. That is another way of saying that the world we observe exists as we perceive it, and that reality can be described reasonably accurately. Those who are not willing to accept evidence must live in an ephemeral world of constantly changing shapes and shadows, because there is no reason to think that anything they believe has any substance.

Asking for evidence for a belief is a powerful way to separate sense from nonsense, which is something politicians should consider. For example, a number of Republican governors and members of Congress believe that reducing income and corporate taxes stimulates the economy. They cannot point to any example of that actually happening, but they believe it with the fervor of a religious doctrine. They say it so often that many of the voters must also believe it, even though there is no evidence for its truth, because these guys get elected and reelected. Kansas provides an excellent example of a state that re-elected a governor even though every part of his tax-cutting program failed. To find the funds needed to balance the state budget, the legislature has been gutting funding for education.

When people do not need evidence to support their most cherished opinions, we should not be surprised if they reject ideas that conflict with those opinions, regardless of how much evidence is advanced for them. After all, what good is evidence that conflicts with what you "know" to be true? It must have been conjured up by people with ulterior motives; people who do not know or who choose not to know the "real truth."

That is the position taken by Creationists who refuse to accept the results of any study that are not consistent with their literal interpretation of the Bible. They believe things that are not physically possible. For example, there is not enough water on the planet for a flood of the magnitude described in the story about Noah. And the Creationists never tell us where all that water went after the flood waters subsided. And what about Joshua commanding the Sun to stop moving across the sky? That would require that the Earth be in the center of the solar system and that the Sun revolves around it. No sane person believes that.

Creationists believe the most fantastical things, but refuse to believe the results of carefully devised experiments, because they feel, rightly, that if their beliefs are not literally true, then they (their beliefs) have no more credence than the beliefs of those in other religions. So the certainty they seem to need is not possible unless the Bible is inerrant.

Their version of reality does not match the real world, but as far as they are concerned that is the real world's problem.

The majority of complaints about regulations pertain to pollution of the environment. It costs money to prevent industrial emissions from polluting the water we drink and the air we breathe, and in principle, society, represented by the government, could provide the financial incentives industry would need to adapt its manufacturing systems so that pollutants are not emitted. But that option amounts to industry blackmailing society.

Are we willing to accept the idea that industry has the right to pollute? When we pay industry for the right to breathe clean air we are admitting that industry "owns" the air. Do we really believe that?

Another way to think about the subject is that industry could say to society, allow us to pollute the air and we will be able to provide the goods you want at a low price. That is another example of blackmail. If we allowed industry to pollute the air we breathe so we can buy products more cheaply, we would be admitting that economics trumps health. Do we really believe that?

Do we want to live in a country in which industrial operations are not subject to any constraints? Are we willing to concede that restrictions on industrial operations that preserve the health of the population are intrusive to the point of being burdensome, so society must pay a price for the privilege of breathing clean air?

Anyone who objects to the recommendations suggested by scientific studies of the environment that are intended to protect the health of the public is answering those questions affirmatively. They are saying that public health considerations are not as important as allowing a corporation to post ever increasing profits each quarter.

Now consider the folks who deny that climate change due to global warming is occurring, and who insist that even if it is, the rise in temperature is not caused by society's use of fossil fuels. Some of them claim that the graphs of rising temperatures measured over the last 200 years mean nothing, because temperatures have always fluctuated, so there is no reason to think that human activities are causing the increase. Evidence that the current fluctuation's rate of growth greatly exceeds any previous ones means nothing to them. Other "deniers" claim that the temperature data have been tampered with statistically to show what the scientists want us all to believe. They have no evidence to support that claim, but evidence is not necessary when a person needs to believe something.

When you will not accept evidence that conflicts with your beliefs, you are saying that the world functions the way you want it to function, and everyone else is wrong. That is how children react to unwelcome information. But how many conflicting viewpoints can exist at any one time, and all of them be true? How can we decide which should be discarded? The only way to avoid this kind of chaotic situation is to use evidence to decide. You will be happy if the evidence supports your opinion, but if it does not, how many people are willing to accept that they may be wrong?

Regardless of expert opinions, we take one side of an issue or the other in a political framework. That is why some people can accept a report (e.g. a UN report on climate change) and others can reject it. When one side of a dispute accepts the advice of experts and the other does not, it seems clear that the accuracy or objectivity of the report is not the point; the kind of life we will lead if a certain decision is made is the point.

Climate deniers are locked into a system that makes no sense. Call it market fundamentalism, to ensure that we do not forget that it involves a quasi-religious set of beliefs. Basic to this set of beliefs is the idea that individuals interact in ways that automatically satisfy their needs, thereby producing a balance between what people want (demands) and what they are offered (supply). Going further, these supposedly self-correcting activities represent the best possible economic system because there are not constraints on peoples' choices. Everyone is free to do what they please – buy and sell, or not buy and sell.

All of that sounds fine, in principle, but depending on this kind of market activities ignores something important, which is "market failures." That is, markets do not always work the way they are supposed to. The negative effects on common resources are the best example of that statement. The effects of excessive use of pesticides, the depletion of the planet's ozone layer due to the leakage of the chemicals once used in used in air conditioning systems, and climate change due to the global warming caused by carbon emissions are all examples of situations for which the "markets" had no solution.

The pollution problems we face today would be considered minimal compared to what would be commonplace without the regulations imposed by governments to increase the cost of certain harmful products. No "naturally occurring" market corrections were observed in the past to deal with the problems associated with harming common resources such as air and water.

That is why I say that climate deniers are locked into a system that makes no sense. Objections to restrictions on economic activity in the name of 'freedom" of people to allow markets to deal with problems leads to disaster.

An Anthropologist once wrote that the practice of magic in primitive societies is useful to resolve disputes because casting a spell on someone is less disruptive than the use of a spear. In an analogous sense, legislation restricting some activities (and enforcement of the legislation) can prevent some problems from occurring, problems which would otherwise require extensive local or political negotiations to resolve, and which would end up satisfying no one.

With regard to climate change, we are at the point at which we desperately need to impose restrictions to prevent activities that have the potential to destroy the society we live in.

Communicating scientific findings to the public has always been a problem, because many people refuse to accept the consensus position of experts in a field. Climate Change is a good example. Quite often the communication problem is not that people do not understand the material, because it is possible to convey the gist of a subject in terms the average person will understand. The problem is that some people just refuse to accept what is presented. And a low level of scientific literacy is not always the cause of the refusal.

For example, studies have been done which asked people to rate the threat of climate change on a scale of one to ten. Then the rating was correlated with their level of scientific literacy. The result was fascinating, because higher scientific literacy correlated strongly with both a high and a low rating of the threat. That is, some people with high scientific literacy are found to rate the threat very high and others with the same degree of scientific literacy rate the threat very low. The conclusion drawn was that people do not use scientific literacy to evaluate what is presented to them. Instead, they use what is presented to them to reinforce their existing attitudes.

This tendency is called "Confirmation Bias," that is, people use information to confirm their prior beliefs, regardless of what that information actually indicates.

The psychologists who did the study concluded that Americans fall into two basic categories. Members of one group, who have what they call a "communitarian" set of values, are generally suspicious of industry and believe that government regulations are warranted whenever they think that proposals may be dangerous. Pollution of the environment is one example. On the other hand, people with a more conservative set of values feel that government interference with their lives or with how society functions is not warranted. They object every time the E.P.A. publishes new air quality rules, saying the rules will destroy the domestic coal industry.

The subject of climate change seems to delineate these two groups very well, so when we argue about climate change, we are really arguing about who we are, and which group we belong to. Our beliefs are influenced by membership in groups. Someone characterized this tendency by saying that all of us are still in high school. We have a need to fit in, and that need can be strong enough to trump logic and reason.

So, for many people, education has no effect on their core beliefs. They may realize that the planet is warming, but they will not concede that the cause is due to human activities, perhaps because they were raised to be suspicious of anything that might imply the need for government regulations. It seems that when faced with a decision about a scientific topic, what they decide has little to do with what they know about the particular topic.

The difference between this "conservative" group and those who accept the opinions of experts, is profound. Those who accept scientific findings differ from the "conservatives" because they accept things that work. This is not necessarily due to Confirmation Bias because they recognize that science has a long track record of getting things right, which is why our technological society functions as well as it does. If accepting what works is a bias, then it is one which benefits society in the long run.

Of course, the same kind of thinking that results in technological progress is used to answer questions about more abstract topics, one of which is climate change. If you have confidence in statements about how the internal combustion engine in your car works, you should have confidence in statements made about how carbon dioxide affects global warming. The reason is that although they are different, both topics are based on some basic principles of heat transfer. If you cannot see the forest for the trees, you did not benefit from your education.

www.ingramcontent.com/pod-product-compliance
Lightning Source LLC
Chambersburg PA
CBHW030838180526
45163CB00004B/1365